Advanced Structured Materials

Volume 121

Common engineering materials reach in many applications their limits and new developments are required to fulfil increasing demands on engineering materials. The performance of materials can be increased by combining different materials to achieve better properties than a single constituent or by shaping the material or constituents in a specific structure. The interaction between material and structure may arise on different length scales, such as micro-, meso- or macroscale, and offers possible applications in quite diverse fields.

This book series addresses the fundamental relationship between materials and their structure on the overall properties (e.g. mechanical, thermal, chemical or magnetic etc) and applications.

The topics of *Advanced Structured Materials* include but are not limited to

- classical fibre-reinforced composites (e.g. glass, carbon or Aramid reinforced plastics)
- metal matrix composites (MMCs)
- micro porous composites
- micro channel materials
- multilayered materials
- cellular materials (e.g., metallic or polymer foams, sponges, hollow sphere structures)
- porous materials
- truss structures
- nanocomposite materials
- biomaterials
- nanoporous metals
- concrete
- coated materials
- smart materials

Advanced Structured Materials is indexed in Google Scholar and Scopus.

More information about this series at http://www.springer.com/series/8611

Holm Altenbach · Michael Brünig ·
Zbigniew L. Kowalewski
Editors

Plasticity, Damage and Fracture in Advanced Materials

Editors
Holm Altenbach
Lehrstuhl Technische Mechanik
Institut für Mechanik
Fakultät für Maschinenbau
Otto-von-Guericke-Universität Magdeburg
Magdeburg, Germany

Michael Brünig
Neubiberg, Bayern, Germany

Zbigniew L. Kowalewski
Warsaw, Poland

ISSN 1869-8433 ISSN 1869-8441 (electronic)
Advanced Structured Materials
ISBN 978-3-030-34853-3 ISBN 978-3-030-34851-9 (eBook)
https://doi.org/10.1007/978-3-030-34851-9

This Springer imprint is published by the registered company Springer Nature Switzerland AG
The registered company address is: Gewerbestrasse 11, 6330 Cham, Switzerland

Preface

This book is a collection of research papers devoted to modeling, experiments and numerical simulations of plasticity, damage and fracture in advanced materials. Currently, various approaches exist based on phenomenological, mechanism-based or physically motivated aspects. In this sense, this book can be seen as some kind of a state of the art.

Accurate modeling of inelastic behavior of advanced materials is an essential point for the solution of boundary-value problems in various engineering disciplines. For example, microscopic defects lead to reduction in strength of materials and can shorten the service and lifetime of engineering structures. Hence, a main aspect of these models is to provide realistic information on the stress distributions within elements of these materials or assessment of safety factors against failure.

During the last years important progress can be seen in the experimental practice delivering important information on inelastic, often localized deformation patterns as well as evolution of damage in interaction with the micro-structure of materials. Great efforts have been made to propose more physically based material models to predict damage and failure in advanced materials or engineering structures under general loading conditions. At the same time, various research fields in solid mechanics and, as a special case, modeling and simulation of behavior of advanced materials has evolved due to development of multi-scale approaches. However, the application of multi-scale models to numerically analyze structures subjected to complex loading scenarios is still at an early stage. Research groups from different countries have proposed promising approaches and some of them are discussed in the present book.

The aim of the book is not only to consolidate the advances in the fields of plasticity, damage and fracture but also to provide a forum to discuss new trends in inelastic material research. New models at several length scales applied to nonlinear and heterogeneous materials have also been emphasized from various related disciplines such as metal physics, micro-mechanics as well as mathematical and computational mechanics.

We would like to express our sincere appreciation to all the authors and the representatives of Springer, who made this volume and the book series possible. We delightedly acknowledge Dr. Christoph Baumann (Editorial Director, Springer

Publisher) for initiating the Advanced Structured Materials Series and this book project. In addition, we have to thank Dr. Mayra Castro (Senior Editor Research Publishing – Books Interdisciplinary Applied Science) and Mr. Ashok Arumairaj (Production Administrator) giving us the final support.

Last but not least, the first editor has to acknowledge the Fundacja na rzecz Nauki Polskiej (Fundation for Polish Science) allowing to finalize this book at the Politechnika Lubelska (host: Prof.dr.hab.inż. Tomasz Sadowski, dr.h.c.) with the help of the Alexander von Humboldt Polish Honorary Research Fellowship.

Magdeburg/Lublin, Neubiberg, Warszawa
August 2019

Holm Altenbach
Michael Brünig
Zbigniew L. Kowalewski

Contents

List of Contributors

Sergei M. Aizikovich
Don State Technical University, 1 Gagarin Sq., Rostov-on-Don, 344000, Russia,
e-mail: saizikovich@gmail.com

Ahmad Alajami
Institute of Applied Mechanics, RWTH Aachen University, Mies-van-der-Rohe-Str. 1 D-52074 Aachen, Germany,
e-mail: ahmad.alajami@rwth-aachen.de

Michael Brünig
Institut für Mechanik und Statik, Universität der Bundeswehr München, Werner-Heisenberg-Weg 39, 85577 Neubiberg, Germany,
e-mail: michael.bruenig@unibw.de

Emrah Demirci
Loughborough University, UK,
e-mail: e.demirci@lboro.ac.uk

Sebastian Dieck
Otto von Guericke University Magdeburg, Institute of Materials and Joining Technology, Universitätsplatz 2, 39106 Magdeburg, Germany,
e-mail: sebastian.dieck@ovgu.de

Ekaterina G. Drogan
Don State Technical University, 1 Gagarin Sq., Rostov-on-Don, 344000, Russia,
e-mail: ekaterina.drogan@gmail.com

Martin Ecke
Otto von Guericke University Magdeburg, Institute of Materials and Joining Technology, Universitätsplatz 2, 39106 Magdeburg, Germany,
e-mail: martin.ecke@ovgu.de

Halina Egner
Institute of Applied Mechanics, Faculty of Mechanical Engineering, Cracow University of Technology, Al. Jana Pawła II 37, 31-864 Kraków, Poland,
e-mail: halina.egner@pk.edu.pl

Władysław Egner
Institute of Applied Mechanics, Faculty of Mechanical Engineering, Cracow University of Technology, Al. Jana Pawła II 37, 31-864 Kraków, Poland,
e-mail: wladyslaw.egner@pk.edu.pl

Sebastian Fritsch
Technical University Chemnitz, Faculty of Mechanical Engineering, Professorship Materials Science, Erfenschlager Straße 73, 09125 Chemnitz, Germany,
e-mail: sebastian.fritsch@mb.tu-chemnitz.de

Steffen Gerke
Institut für Mechanik und Statik, Universität der Bundeswehr München, Werner-Heisenberg-Weg 39, 85577 Neubiberg, Germany,
e-mail: steffen.gerke@unibw.de

Thorsten Halle
Otto von Guericke University Magdeburg, Institute of Materials and Joining Technology, Universitätsplatz 2, 39106 Magdeburg, Germany,
e-mail: thorsten.halle@ovgu.de

Szymon Hernik
Institute of Applied Mechanics, Faculty of Mechanical Engineering, Cracow University of Technology, Al. Jana Pawła II 37, 31-864 Kraków, Poland,
e-mail: hernik@mech.pk.edu.pl

Sebastian Hütter
Otto von Guericke University Magdeburg, Institute of Materials and Joining Technology, Universitätsplatz 2, 39106 Magdeburg, Germany,
e-mail: sebastian.huetter@ovgu.de

Michael Kaliske
Institute for Structural Analysis, TU Dresden, Germany,
e-mail: Michael.Kaliske@tu-dresden.de

Valeriy L. Khavin
Department of Continuum Mechanics and Strength of Materials, National Technical University "Kharkiv Polytechnic Institute", 61002 Kharkiv, Ukraine,
e-mail: vkhavin@kpi.kharkov.ua

Roman V. Karotkiyan
Don State Technical University, 1 Gagarin Sq., Rostov-on-Don, 344000, Russia,
e-mail: valker94@gmail.com

Evgeniy A. Kislyakov
Don State Technical University, 1 Gagarin Sq., Rostov-on-Don, 344000, Russia,
e-mail: evgenka_95@mail.ru

Zbigniew L. Kowalewski
Institute of Fundamental Technological Research, ul. Pawińskiego 5B, 02-106 Warsaw, Poland,
e-mail: zkowalew@ippt.pan.pl

Manja Krüger
Forschungszentrum Jülich GmbH, Institute of Energy and Climate Research (IEK), Microstructure and Properties of Materials (IEK-2), 52425 Juelich, Germany,
e-mail: manja.krueger@ovgu.de

Dominik Kukla
Institute of Fundamental Technological Research, ul. Pawińskiego 5B, 02-106 Warsaw, Poland,
e-mail: dkukla@ippt.pan.pl

Yujun Li
Northwestern Polytechnical University, 127 Youyi W Rd, Beilin Qu, Xian Shi, Shaanxi Sheng, China,
e-mail: li.yujun@nwpu.edu.cn

Katarzyna Makowska
Motor Transport Institute ul. Jagiellonska 80, 03-301 Warsaw, Poland,
e-mail: katarzyna.makowska@its.waw.pl

Stanislav Yu. Maksyukov
Don State Technical University, 1 Gagarin Sq., Rostov-on-Don, 344000, Russia,
e-mail: kafstom2.rostgmu@yandex.ru

Oliver Michael
Otto von Guericke University Magdeburg, Institute of Materials and Joining Technology, Universitätsplatz 2, 39106 Magdeburg, Germany,
e-mail: oliver.michael@st.ovgu.de

Stanisław Mroziński
UTP University of Science and Technology, Faculty of Mechanical Engineering, Bydgoszcz, Poland,
e-mail: stanislaw.mrozinski@utp.edu.pl

Konstantin Naumenko
Institut für Mechanik, Otto-von-Guericke-Universität Magdeburg, 39106 Magdeburg, Germany,
e-mail: konstantin.naumenko@ovgu.de

Andrey L. Nikolaev
Don State Technical University, 1 Gagarin Sq., Rostov-on-Don, 344000, Russia,
e-mail: andreynicolaev@eurosites.ru

Sergiy Yu. Pogorilov
Department of Continuum Mechanics and Strength of Materials, National Technical University "Kharkiv Polytechnic Institute", 61002 Kharkiv, Ukraine,
e-mail: ark95@ukr.net

Paul Rosemann
Otto von Guericke University Magdeburg, Institute of Materials and Joining Technology, Universitätsplatz 2, 39106 Magdeburg, Germany,
e-mail: paul.rosemann@ovgu.de

Evgeniy V. Sadyrin
Don State Technical University, 1 Gagarin Sq., Rostov-on-Don, 344000, Russia,
e-mail: ghostwoode@gmail.com

Kyrill Yu. Schastlivets
Department of Computer Modeling of Processes and Systems, National Technical University "Kharkiv Polytechnic Institute", 61002 Kharkiv, Ukraine,
e-mail: sti66@mail.ru

Vadim V. Silberschmidt
Loughborough University, UK,
e-mail: v.silberschmidt@lboro.ac.uk

Jaan-Willem Simon
Institute of Applied Mechanics, RWTH Aachen University, Mies-van-der-Rohe-Str. 1 D-52074 Aachen, Germany,
e-mail: jaan.simon@rwth-aachen.de

Emrah Sozumert
Loughborough University, UK,
e-mail: e.sozumert@lboro.ac.uk

Christian Steinke
Institute for Structural Analysis, TU Dresden, Germany,
e-mail: Christian.Steinke@tu-dresden.de

Piotr Sulich
Institute of Applied Mechanics, Faculty of Mechanical Engineering, Cracow University of Technology, Al. Jana Pawła II 37, 31-864 Kraków, Poland
e-mail: piotrjansulich@gmail.com

Michael V. Swain
Don State Technical University, 1 Gagarin Sq., Rostov-on-Don, 344000, Russia,
e-mail: michael.swain@sydney.edu.au

Tadeusz Szymczak
Motor Transport Institute ul. Jagiellonska 80, 03-301 Warsaw, Poland,
e-mail: tadeusz.szymczak@its.waw.pl

Aneta Ustrzycka
Institute of Fundamental Technological Research, ul. Pawińskiego 5B, 02-106 Warsaw, Poland,
e-mail: austrzyc@ippt.pan.pl

Martin Franz-Xaver Wagner
Technical University Chemnitz, Faculty of Mechanical Engineering, Professorship Materials Science, Erfenschlager Straße 73, 09125 Chemnitz, Germany,
e-mail: martin.wagner@mb.tu-chemnitz.de

Markus Wilke
Otto von Guericke University Magdeburg, Institute of Materials and Joining Technology, Universitätsplatz 2, 39106 Magdeburg, Germany,
e-mail: markus.wilke@ovgu.de

Anna Wiśniewska
Institute of Applied Mechanics, Faculty of Mechanical Engineering, Cracow University of Technology, Al. Jana Pawła II 37, 31-864 Kraków, Poland
e-mail: anna.wisniewska1@pk.edu.pl

Diana V. Yogina
Rostov State Medical University, 29 Nakhichevansky Lane, Rostov-on-Don, 344022, Russia,
e-mail: dianaturbina@mail.ru

Moritz Zistl
Institut für Mechanik und Statik, Universität der Bundeswehr München, Werner-Heisenberg-Weg 39, 85577 Neubiberg, Germany,
e-mail: moritz.zistl@unibw.de

Imadeddin Zreid
Institute for Structural Analysis, TU Dresden, Germany,
e-mail: Imadeddin.Zreid@tu-dresden.de

Chapter 1
Continuum Modelling of the Anisotropic Elastic-Plastic In-Plane Behavior of Paper in Small and Large Strains Regimes

Ahmad Alajami, Yujun Li, and Jaan-Willem Simon

Abstract Laminated paperboard is one of the main materials used in packaging industry. The manufacturing process of such material leads to a special distribution of the fibers. Such distribution gives the material an anisotropic elastic-plastic behavior. The formulation of the final packages requires different complicated industrial processes such as creasing and folding. During these processes, the material undergoes large deformations, which requires analyzing the material in the large strains regime. In the following chapter, the analysis of the anisotropic elastic-plastic behavior of the material will be introduced in the small strains regime, and then extended to the large strains regime. The punch test will be presented at the end of this chapter and will be used for the validation of the model as it contains complex loadings.

Key words: Paper · Anisotropic plasticity · Large strains · Isotropic hardening

1.1 Introduction

The manufacturing process of laminated paperboard introduces a special fiber distribution. Due to this distribution, three principal directions can be distinguished; Machine Direction (MD), Cross Direction (CD) and Thickness Direction (ZD). Most of the fibers are oriented along the MD direction, which yields to a higher stiffness than in the other directions, while only few fibers are located along the thickness direction (ZD). Formulation of the paperboard packages requires a creasing process,

---

Ahmad Alajami · Jaan-Willem Simon
Institute of Applied Mechanics, RWTH Aachen University, Mies-van-der-Rohe-Str. 1 D-52074 Aachen, Germany,
e-mail: ahmad.alajami@rwth-aachen.de, jaan.simon@rwth-aachen.de

Yujun Li at Northwestern Polytechnical University, 127 Youyi W Rd, Beilin Qu, Xian Shi, Shaanxi Sheng, China,
e-mail: li.yujun@nwpu.edu.cn

H. Altenbach et al. (eds.), *Plasticity, Damage and Fracture in Advanced Materials*, Advanced Structured Materials 121,
https://doi.org/10.1007/978-3-030-34851-9_1

in which, a creasing rule is pushed against the paperboard introducing a locally deformed zone. Then, a subsequent folding process is implemented to form the final package. During this process, the outer layers undergo in-plane tensile loadings, and the inner layers undergo in-plane compressive loadings, and the opposite happens during the folding process. To analyze the anisotropic behavior of the material during these processes, a continuum model is introduced to describe the in-plane and out-of-plane behavior of the material. In the literature, different material models were introduced to describe the elastic-plastic behavior of paper. Xia et al (2002) introduced a multi-surface yield function to determine the initiation of plasticity. This function was accompanied with a non-linear hardening functions to describe the anisotropic strain hardening observed in paper. This model was further developed by Borgqvist et al (2014) by including distortional hardening. In this model, additional set of coupling parameters were introduced. However, the determination of such parameters requires an extensive experimental effort. Other approaches included the analysis of the creasing and folding processes based on the simple yield criterion proposed by Hill (1948) which has been implemented to describe the onset of yield and the plastic flow as shown by Beex and Peerlings (2012); Huang et al (2014).

In the following chapter, a continuum model for describing in-plane behavior of paper will be presented. To account for the material's anisotropy, the concept of structural tensors is applied. Further, the yield criterion developed by Li et al (2016) based on the formulation of Xia et al (2002) is adopted and extended to adapt to the large strains regime.

1.2 Continuum Model for Small Strains

The in-plane model presented in this section has been developed by Li et al (2016) for the small strains regime. The current model can represent the elastic-plastic anisotropy observed in experiments To model the elastic plastic behavior an additive decomposition of the total strain tensor $\boldsymbol{\varepsilon}$ into a reversible elastic part $\boldsymbol{\varepsilon}^e$, and an irreversible plastic part $\boldsymbol{\varepsilon}^p$ is utilized:

$$\boldsymbol{\varepsilon} = \boldsymbol{\varepsilon}^e + \boldsymbol{\varepsilon}^p \tag{1.1}$$

1.2.1 Thermodynamic Relations

The starting point of the derivation of the constitutive equations is given by the second law of thermodynamics. For isothermal processes, the Clausius-Duhem inequality needs to be satisfied

$$-\dot{\psi} + \boldsymbol{\sigma} : \dot{\boldsymbol{\varepsilon}} \geq 0, \tag{1.2}$$

where ψ is the Helmholtz free energy, $\boldsymbol{\sigma}$ is the Cauchy stress tensor, and $\dot{\boldsymbol{\varepsilon}}$ is the strain rate tensor. The free energy can be split into an elastic part ψ^e and a plastic part ψ^p:

$$\psi = \psi^e\left(\boldsymbol{\varepsilon}^e, \boldsymbol{m}_\alpha\right) + \psi^p\left(\kappa\right), \tag{1.3}$$

To describe the directional dependency of the material, the elastic part of the free energy is formulated based on the concept of invariants formed by the structural tensors

$$\boldsymbol{m}_\alpha = \boldsymbol{n}_\alpha \otimes \boldsymbol{n}_\alpha, \tag{1.4}$$

where $\boldsymbol{n}_\alpha$ $(\alpha = 1, 2)$ represent the unit vectors of the material directions, namely MD and CD of paper. Orthotropic plasticity is accounted for by applying an orthotropic yield function. However, the strain hardening is assumed to be isotropic and hence the plastic part of the free energy is a function of the internal variable κ which accounts for the irreversible effects representing the isotropic hardening. By substituting (1.3) into (1.2), the dissipation inequality becomes

$$\left(\boldsymbol{\sigma} - \frac{\partial \psi^e}{\partial \boldsymbol{\varepsilon}^e}\right) : \dot{\boldsymbol{\varepsilon}}^e + \boldsymbol{\sigma} : \dot{\boldsymbol{\varepsilon}}^p - \frac{\partial \psi^p}{\partial \kappa}\dot{\kappa} \geq 0. \tag{1.5}$$

This inequality must be fulfilled for arbitrary thermodynamic processes, i.e. for arbitrary $\dot{\boldsymbol{\varepsilon}}^e$, $\dot{\boldsymbol{\varepsilon}}^p$, and $\dot{\kappa}$. Therefore, the Cauchy stress can be obtained by

$$\boldsymbol{\sigma} = \frac{\partial \psi^e}{\partial \boldsymbol{\varepsilon}^e}, \tag{1.6}$$

and the remaining part of the Clausius–Duhem inequality is given as

$$\boldsymbol{\sigma} : \dot{\boldsymbol{\varepsilon}}^p - R\,\dot{\kappa} \geq 0, \tag{1.7}$$

where the following conjugated quantity to the internal variable κ has been introduced

$$R = \frac{\partial \psi^p}{\partial \kappa}. \tag{1.8}$$

1.2.2 Elasticity

The linear elastic response of paper in the in-plane direction is represented by adopting the orthotropic model of Reese et al (2001). The elastic part of the strain energy function takes the following form:

$$\begin{aligned}\psi^e = {} & K_1^{\mathrm{iso}} {I_1}^2 + K_2^{\mathrm{iso}} I_2 + K_1^{\mathrm{ani1}} {I_4}^2 + K_2^{\mathrm{ani1}} \left(-I_4 + I_5\right) + K_1^{\mathrm{ani2}} {I_6}^2 \\ & + K_2^{\mathrm{ani2}} \left(-I_6 + I_7\right) + K^{\mathrm{coup1}} I_1 I_4 + K^{\mathrm{coup2}} I_1 I_6 + K^{\mathrm{coup\ ani}} I_4 I_6,\end{aligned} \tag{1.9}$$

where K_1^{iso}, K_2^{iso}, K_1^{ani1}, K_2^{ani1}, K_1^{ani2}, K_2^{ani2}, K^{coup1}, K^{coup2}, $K^{\mathrm{coup\ ani}}$ are material parameters which have to be fitted to experiments. Their physical meaning will be discussed in the calibration Subsect. 1.4.1. In addition, the invariants, I_i $(i = 1, \ldots, 7)$, are defined as:

$$
\begin{aligned}
&I_1 := \operatorname{tr}\boldsymbol{\varepsilon}^e,\ I_2 := \frac{1}{2}\left(I_1{}^2 - \operatorname{tr}(\boldsymbol{\varepsilon}^e)^2\right),\ I_3 := \det \boldsymbol{\varepsilon}^e,\\
&I_4 := \operatorname{tr}\left(\boldsymbol{\varepsilon}^e \boldsymbol{m}_1\right) = \boldsymbol{\varepsilon}^e : \boldsymbol{m}_1,\ I_5 := \operatorname{tr}\left((\boldsymbol{\varepsilon}^e)^2 \boldsymbol{m}_1\right) = (\boldsymbol{\varepsilon}^e)^2 : \boldsymbol{m}_1,\\
&I_6 := \operatorname{tr}\left(\boldsymbol{\varepsilon}^e \boldsymbol{m}_2\right) = \boldsymbol{\varepsilon}^e : \boldsymbol{m}_2,\ I_7 := \operatorname{tr}\left((\boldsymbol{\varepsilon}^e)^2 \boldsymbol{m}_2\right) = (\boldsymbol{\varepsilon}^e)^2 : \boldsymbol{m}_2.
\end{aligned}
\tag{1.10}
$$

Noteworthy, similar expressions have also been given by Naumenko and Altenbach (2016) for modeling of creep.

By substituting (1.9) into (1.6), the Cauchy stress tensor can be defined as:

$$
\sigma = \frac{\partial \psi^e}{\partial \boldsymbol{\varepsilon}^e} = \mathscr{L} : \boldsymbol{\varepsilon}^e \tag{1.11}
$$

where the fourth order material tensor $\mathscr{L}$ can be calculated by

$$
\begin{aligned}
\mathscr{L} =\, & 2K_1^{\text{iso}}\mathbf{I}\otimes\mathbf{I} + K_2^{\text{iso}}\left(\mathbf{I}\otimes\mathbf{I} - \mathscr{I}\right) + 2K_1^{\text{ani1}}\boldsymbol{m}_1\otimes\boldsymbol{m}_1\\
&+ K_2^{\text{ani1}}\mathscr{E}_1 + 2K_1^{\text{ani2}}\boldsymbol{m}_2\otimes\boldsymbol{m}_2 + K_2^{\text{ani2}}\mathscr{E}_2 + K^{\text{coup1}}\left(\mathbf{I}\otimes\boldsymbol{m}_1 + \boldsymbol{m}_1\otimes\mathbf{I}\right)\\
&+ K^{\text{coup2}}\left(\mathbf{I}\otimes\boldsymbol{m}_2 + \boldsymbol{m}_2\otimes\mathbf{I}\right) + K^{\text{coup ani}}\left(\boldsymbol{m}_1\otimes\boldsymbol{m}_2 + \boldsymbol{m}_2\otimes\boldsymbol{m}_1\right).
\end{aligned}
\tag{1.12}
$$

The coefficients of the fourth-order tensors $\mathscr{E}_1$ and $\mathscr{E}_2$ are given by

$$
(\mathscr{E}_1)_{ijkl} = (\boldsymbol{m}_1)_{ik}\delta_{jl} + (\boldsymbol{m}_1)_{jk}\delta_{il} \text{ and } (\mathscr{E}_2)_{ijkl} = (\boldsymbol{m}_2)_{ik}\delta_{jl} + (\boldsymbol{m}_2)_{jk}\delta_{il}, \tag{1.13}
$$

and $\mathbf{I}$ and $\mathscr{I}$ denote the second-order and fourth-order identity tensor, respectively. The stiffness tensor in Voigt notation can be further reduced to fit for the in-plane loadings

$$
\hat{\mathscr{L}}_{ip} = \left[\begin{array}{c|c|c}
2K_1^{\text{ani1}} & -K_2^{\text{iso}} + K^{\text{coup ani}} & 0\\
\hline
-K_2^{\text{iso}} + K^{\text{coup ani}} & 2K_1^{\text{ani2}} & 0\\
\hline
0 & 0 & K_2^{\text{iso}}
\end{array}\right]. \tag{1.14}
$$

To get better physical understanding of the previous parameters, the in-plane stiffness tensor can be compared to the classical formation of the in-plane orthotropic elasticity tensor

$$
\hat{\mathscr{C}}_{ip} = \frac{1}{1-\nu_{12}\nu_{21}}\left[\begin{array}{c|c|c}
E_{11} & \nu_{21}E_{11} & 0\\
\hline
\nu_{12}E_{22} & E_{22} & 0\\
\hline
0 & 0 & (1-\nu_{12}\nu_{21})\,G_{12}
\end{array}\right], \tag{1.15}
$$

From this comparison, it can be concluded that the parameters K_1^{ani1} and K_1^{ani2} represent the elastic modulus in the MD and CD directions. The Poisson's effect can be described with the parameter $K^{\text{coup ani}}$, while the shear modulus can be represented by K_2^{iso}.

1.2.3 Plasticity

The previous work of Xia et al (2002) is modified to describe the plastic deformation in paper. The plastic part of the strain energy function ψ^p given in (1.3) accounting to the isotropic hardening can be described as

$$\psi^p = Q\left(\kappa + \frac{\mathrm{e}^{-\beta\kappa}}{\beta}\right) \tag{1.16}$$

where β and Q are material parameters, which can be fitted based on in-plane tensile tests. The internal variable κ is the accumulated plastic strain in the isotropic hardening contribution. By implementing (1.8), the thermodynamic force associated with isotropic hardening is, then, given by

$$R = Q\left(1 - \mathrm{e}^{-\beta\kappa}\right). \tag{1.17}$$

The original multi-surface based yield criterion is modified to be

$$\Phi\left(\boldsymbol{\sigma}, R\right) = \left\{\sum_{\alpha=1}^{6}\left\{x_\alpha\left[\frac{\boldsymbol{\sigma} : \mathbf{M}^\alpha}{r_\alpha^0}\right]^{2k}\right\}\right\}^{\frac{1}{2k}} - (\sigma_0 + R), \tag{1.18}$$

where x_α is a switch to determine whether the current stress activates the αth yield plane, and is defined as

$$x_\alpha = \begin{cases} 1 \text{ if } \boldsymbol{\sigma} : \mathbf{M}^\alpha > 0, \\ 0 \qquad \text{otherwise.} \end{cases} \tag{1.19}$$

k is an integer which is greater than or equal to 1. This parameter is used to smooth the corners between the adjacent subsurfaces, and can be determined from biaxial tests. Increasing the value of k leads to sharper corner and decreasing its value leads to smoother corners. The parameter σ_0 represents the reference initial yield stress. To account for the anisotropy observed in the initial yield stress in different directions, the parameters r_α^0 are implemented. These parameters represent the ratio of distance between the origin and the αth yield plane to the reference value σ_0, and can be derived by implementing the same procedure implemented by Xia et al (2002)

$$
\begin{aligned}
r_1^0 &= \left(\frac{\Omega_2}{\Omega_1}\right)^{\frac{1}{2k}}, \quad r_2^0 = \left(\frac{\Omega_4}{\Omega_3}\right)^{\frac{1}{2k}}, \\
r_3^0 &= R_{\mathrm{CD}}^t M_{22}^3 \left[1 - \left(R_{\mathrm{CD}}^t M_{22}^2\right)^{2k} \frac{\Omega_3}{\Omega_4}\right]^{-\frac{1}{2k}}, \quad r_4^0 = R_{\mathrm{CD}}^t M_{11}^4 \left[1 - \left(R_{\mathrm{MD}}^t M_{11}^1\right)^{2k} \frac{\Omega_1}{\Omega_2}\right]^{-\frac{1}{2k}}, \\
r_5^0 &= r_6^0 = \frac{R_{45^\circ}^t M_{12}^5}{2} \left\{1 - \left[\frac{R_{45^\circ}^t}{2r_1^0}\left(M_{11}^1 + M_{22}^1\right)\right]^{2k} - \left[\frac{R_{45^\circ}^t}{2r_2^0}\left(M_{11}^2 + M_{22}^2\right)\right]^{2k}\right\}^{-\frac{1}{2k}},
\end{aligned}
\tag{1.20}
$$

where

$$
\begin{aligned}
&\text{reference stress:} \quad \sigma_0 = \sigma_{\mathrm{MD}}^{t0}, \\
&\text{MD tension:} \quad R_{\mathrm{MD}}^t = \frac{\sigma_{\mathrm{MD}}^{t0}}{\sigma_0}, \qquad \text{MD compression:} \quad R_{\mathrm{MD}}^c = \frac{\sigma_{\mathrm{MD}}^{c0}}{\sigma_0}, \\
&\text{CD tension:} \quad R_{\mathrm{CD}}^t = \frac{\sigma_{\mathrm{CD}}^{t0}}{\sigma_0}, \qquad \text{CD compression:} \quad R_{\mathrm{CD}}^c = \frac{\sigma_{\mathrm{CD}}^{c0}}{\sigma_0}, \\
&45^\circ \text{ tension:} \quad R_{45^\circ}^t = \frac{\sigma_{45^\circ}^{t0}}{\sigma_0}.
\end{aligned}
\tag{1.21}
$$

and

$$
\begin{aligned}
\Omega_1 &= \left[R_{\mathrm{CD}}^c M_{22}^4\right]^{2k} - \left[R_{\mathrm{MD}}^t M_{11}^4\right]^{2k}, \\
\Omega_2 &= \left[R_{\mathrm{MD}}^t R_{22}^c\right]^{2k} \left[\left(M_{11}^1 M_{22}^4\right)^{2k} - \left(M_{22}^1 M_{11}^4\right)^{2k}\right], \\
\Omega_3 &= \left[R_{\mathrm{MD}}^c M_{11}^3\right]^{2k} - \left[R_{\mathrm{CD}}^t M_{22}^3\right]^{2k}, \\
\Omega_4 &= \left[R_{\mathrm{MD}}^c R_{\mathrm{CD}}^t\right]^{2k} \left[\left(M_{11}^3 M_{22}^2\right)^{2k} - \left(M_{22}^3 M_{11}^2\right)^{2k}\right].
\end{aligned}
\tag{1.22}
$$

A graphical representation of the yield surface illustrating the intersection of the yield surface with the $\sigma_{11} - \sigma_{22}$ plane, in the absence of shear stresses is shown in Fig. 1.1 for three different choices of the parameter $2k$. To describe the non-isochoric plastic deformation, a set of yield plane tensors, $\mathbf{M}^\alpha$, normal to the yield plane, are introduced as

$$
\mathbf{M}^\alpha = M_{11}^\alpha \boldsymbol{m}_1 + M_{22}^\alpha \boldsymbol{m}_2 + M_{12}^\alpha \left(\boldsymbol{n}_1 \otimes \boldsymbol{n}_2 + \boldsymbol{n}_2 \otimes \boldsymbol{n}_1\right), \tag{1.23}
$$

where M_{11}^α, M_{22}^α and M_{12}^α are parameters fitted from experiments. These plane normals determine the nearly constant plastic flow direction. The components of these normals can be fitted from experiments and are given as follows

$$
M_{ij}^1 = \frac{1}{\sqrt{1 + d_{\mathrm{MD}}^2}} \begin{bmatrix} 1 & 0 & 0 \\ 0 & d_{\mathrm{MD}} & 0 \\ 0 & 0 & 0 \end{bmatrix}, \qquad M_{ij}^2 = \frac{1}{\sqrt{1 + d_{\mathrm{CD}}^2}} \begin{bmatrix} d_{\mathrm{CD}} & 0 & 0 \\ 0 & 1 & 0 \\ 0 & 0 & 0 \end{bmatrix}, \tag{1.24}
$$

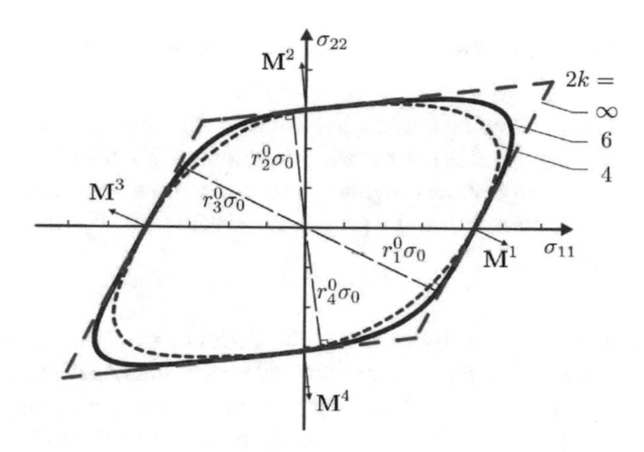

Fig. 1.1 Yield surface for the biaxial loading state

The parameters d_{MD} and d_{CD} represent the ratio between the lateral and axial plastic strains during the in-plane tensile test. To account for the preservation of the plastic deformation in the pure shear mode, the shear yield plane normals $\mathbf{M}^5$ and $\mathbf{M}^6$ are determined as follows:

$$M_{ij}^5 = \frac{1}{\sqrt{2}} \begin{bmatrix} 0 & 1 & 0 \\ 1 & 0 & 0 \\ 0 & 0 & 0 \end{bmatrix}, \qquad M_{ij}^6 = \frac{-1}{\sqrt{2}} \begin{bmatrix} 0 & 1 & 0 \\ 1 & 0 & 0 \\ 0 & 0 & 0 \end{bmatrix}, \tag{1.25}$$

The remaining normals in the negative stress state can be determined to be the opposite of their corresponding yield planes in the positive stress state $\mathbf{M}^3 = -\mathbf{M}^1$ and $\mathbf{M}^4 = -\mathbf{M}^2$. The evolution equations for the plastic strains can be given as

$$\begin{aligned} \dot{\boldsymbol{\varepsilon}}^p &= \dot{\lambda} \boldsymbol{N}, \\ \dot{\kappa} &= \dot{\lambda} \end{aligned} \tag{1.26}$$

where $\dot{\lambda}$ is the plastic multiplier, and $\mathbf{N}$ is defined by

$$\boldsymbol{N} = \frac{\partial \Phi}{\partial \boldsymbol{\sigma}} \bigg/ \left\| \frac{\partial \Phi}{\partial \boldsymbol{\sigma}} \right\|. \tag{1.27}$$

Assuming an associative flow rule, the model equations can now be summarized as follows:

$$\begin{aligned} &\boldsymbol{\sigma} = \mathscr{L} : \boldsymbol{\varepsilon}^e, \\ &\dot{\boldsymbol{\varepsilon}}^p = \dot{\lambda} \boldsymbol{N}, \quad \dot{\kappa} = \dot{\lambda}, \\ &\Phi = \left\{ \sum_{\alpha=1}^{6} \left\{ \chi_\alpha \left[\frac{\boldsymbol{\sigma} : \mathbf{M}^\alpha}{r_\alpha^0} \right]^{2k} \right\} \right\}^{\frac{1}{2k}} - (\sigma_0 + R), \quad R = Q\left(1 - \mathrm{e}^{-\beta\kappa}\right) \\ &\dot{\lambda} \geq 0, \quad \Phi \leq 0, \quad \dot{\lambda}\Phi = 0. \end{aligned} \tag{1.28}$$

1.3 Continuum Model for Large Strains

In the following section, the extension of the in-plane model from the small strains to the large strains regime will be presented. To model elastic-plastic behavior of paper in the large strains regime, the total deformation gradient $\mathbf{F}$ is split into a reversible elastic part, $\mathbf{F}_e$, and a permanent plastic part, $\mathbf{F}_p$, using the classical multiplicative split

$$\mathbf{F} = \mathbf{F}_e\mathbf{F}_p \tag{1.29}$$

The multiplicative split of the deformation gradient requires the introduction of three different configurations; reference, intermediate, and current configurations as shown in Fig. 1.2 Such split introduces multiple strain tensors.

The elastic $\mathbf{C}_e$ and plastic $\mathbf{C}_p$ parts of the right Cauchy-Green tensors are introduced as

$$\mathbf{C} = \mathbf{F}^T\mathbf{F}, \qquad \mathbf{C}_e = \mathbf{F}_e^T\mathbf{F}_e, \qquad \mathbf{C}_p = \mathbf{F}_p^T\mathbf{F}_p, \tag{1.30}$$

Furthermore, three unit vectors $\mathbf{N}_\alpha$ $(\alpha = 1,2,3)$, are introduced in the reference configuration. These vectors are aligned with the main directions in paper; MD, CD and ZD. The vectors are initially orthogonal to each other , i.e. $\mathbf{N}_i \cdot \mathbf{N}_j = \delta_{ij}$. Identity mapping can be assumed between the director vectors in the reference configuration $\mathbf{N}_\alpha$ and the ones in the intermediate configuration according to the anisotropic evolution analysis on paper as in (Borgqvist et al, 2014). Based on these vectors, three second-order structural tensors can be introduced as follows

$$\mathbf{M}_\alpha = \mathbf{N}_\alpha \otimes \mathbf{N}_\alpha, \; \alpha = 1,2,3. \tag{1.31}$$

Due to the orthotropic feature of MD, CD, and ZD in paperboard, one of the structural tensors $\mathbf{M}_\alpha$ can be expressed as a function of the other two based on the following equaiton

$$\mathbf{M}_1 + \mathbf{M}_2 + \mathbf{M}_3 = \mathbf{I} \tag{1.32}$$

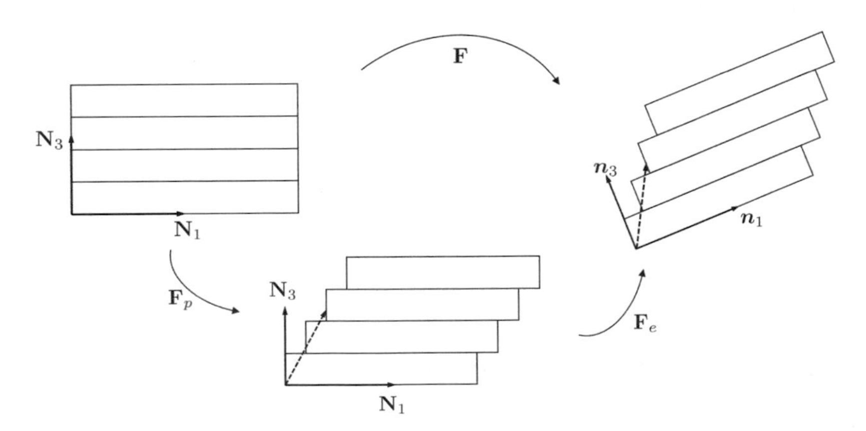

Fig. 1.2 Decomposition of the deformation gradient (modified from Borgqvist et al, 2014)

1.3.1 Thermodynamic Relations

The derivation of the constitutive equations for the large strains regime follows similar approach to the small strains regime. The Clausius-Duhem inequality is modified to take the form

$$-\dot{\psi} + \mathbf{S} : \frac{1}{2}\dot{\mathbf{C}} \geq 0, \tag{1.33}$$

where ψ is the Helmholtz free energy and $\mathbf{S}$ is the second Piola-Kirchhoff stress tensor. The free energy can be split into an elastic part ψ^e and a plastic part ψ^p:

$$\psi = \psi^e \left(\mathbf{C}_e, \mathbf{M}_\alpha\right) + \psi^p \left(\mathbf{A}\right), \tag{1.34}$$

where the elastic part of the free energy is a function of the elastic part of the Cauchy-Green tensor $\mathbf{C}_e$, and the plastic part is a function of a set of internal variables $\mathbf{A}$. By substituting 1.34 into 1.33, the dissipation inequality becomes as follows

$$\left(\mathbf{S} - 2\mathbf{F}_p^{-1}\frac{\partial \psi_e}{\partial \mathbf{C}_e}\mathbf{F}_p^{-T}\right) : \frac{1}{2}\dot{\mathbf{C}} + \mathbf{M} : \mathbf{L}_p - \frac{\partial \psi_p}{\partial \kappa} \cdot \dot{\kappa} \geq 0, \tag{1.35}$$

where $\mathbf{L}_p$ is the plastic deformation rate tensor, defined by

$$\mathbf{L}_p = \dot{\mathbf{F}}_p \mathbf{F}_p{}^{-1}, \tag{1.36}$$

and $\mathbf{\Sigma}$ is the Mandel stress tensor

$$\mathbf{\Sigma} := 2\mathbf{C}_e \frac{\partial \psi_e}{\partial \mathbf{C}_e} \tag{1.37}$$

To satisfy the inequality for any arbitary thermodynamic process, the second Piola-Kirchoff stress tensor should take the following form

$$\mathbf{S} = 2\mathbf{F}_p^{-1}\frac{\partial \psi_e}{\partial \mathbf{C}_e}\mathbf{F}_p^{-T}. \tag{1.38}$$

and the remaining part of the Clausius-Duhem inequality will be reduced to the following form

$$\mathbf{\Sigma} : \mathbf{L}_p - \mathbf{\Theta} \cdot \dot{\mathbf{A}} \geq 0, \tag{1.39}$$

where the conjugated forces θ have been introduced according to

$$\mathbf{\Theta} = \frac{\partial \psi_p}{\partial \mathbf{A}}. \tag{1.40}$$

1.3.2 Elasticity

The behavior of paper under in-plane loadings can be described by a linear relationship. The elastic response of the material can be described by adopting the previous work of Reese et al (2001); Reese (2003)

$$\begin{aligned}\psi^e =& K_2^{\text{iso}}\,(I_2-2I_1+3)+K_1^{\text{ani1}}(I_4-1)^2+K_2^{\text{ani1}}\,(I_5-4I_4+3)+K_1^{\text{ani2}}(I_6-1)^2\\ &+K_2^{\text{ani2}}\,(I_7-4I_6+3)+K^{\text{coup1}}\,(I_1-3)\,(I_4-1)+K^{\text{coup2}}\,(I_1-3)\,(I_6-1)\\ &+K^{\text{coup ani}}\,(I_4-1)\,(I_6-1)\end{aligned} \tag{1.41}$$

where $K_2^{\text{iso}}, K_1^{\text{ani1}}, K_2^{\text{ani1}}, K_1^{\text{ani2}}, K_2^{\text{ani2}}, K^{\text{coup1}}, K^{\text{coup2}}, K^{\text{coup ani}}$ are material parameters which have to be fit to experiments. In addition, the invariants, I_i $(i = 1, \ldots, 7)$, are given as

$$\begin{aligned}&I_1 := \text{tr}\mathbf{C}_e,\; I_2 := \frac{1}{2}\left(I_1^2-\text{tr}\left(\mathbf{C}_e^2\right)\right),\; I_3 := \det\mathbf{C}_e,\\ &I_4 := \text{tr}\,(\mathbf{C}_e\mathbf{M}_1)=\mathbf{C}_e:\mathbf{M}_1,\; I_5 := \text{tr}\left(\mathbf{C}_e^2\mathbf{M}_1\right)=\mathbf{C}_e^2:\mathbf{M}_1,\\ &I_6 := \text{tr}\,(\mathbf{C}_e\mathbf{M}_2)=\mathbf{C}_e:\mathbf{M}_2,\; I_7 := \text{tr}\left(\mathbf{C}_e^2\mathbf{M}_2\right)=\mathbf{C}_e^2:\mathbf{M}_2.\end{aligned} \tag{1.42}$$

By implementing Eq. (1.41), the component $\frac{\partial\psi_e}{\partial\mathbf{C}_e}$ can be expressed as follows

$$\begin{aligned}\frac{\partial\psi_e}{\partial\mathbf{C}_e} =& K_2^{\text{iso}}\,(I_1\mathbf{I}-\mathbf{C}_e-2\mathbf{I})+K^{\text{coup ani}}\,[(I_6-1)\,\mathbf{M}_1+(I_4-1)\,\mathbf{M}_2]\\ &+2K_1^{\text{ani1}}\,(I_4-1)\,\mathbf{M}_1+K_2^{\text{ani1}}\,(-4\mathbf{M}_1+\mathbf{C}_e\mathbf{M}_1+\mathbf{M}_1\mathbf{C}_e)\\ &+2K_1^{\text{ani2}}\,(I_6-1)\,\mathbf{M}_2+K_2^{\text{ani2}}\,(-4\mathbf{M}_2+\mathbf{C}_e\mathbf{M}_2+\mathbf{M}_2\mathbf{C}_e)\\ &+K^{\text{coup1}}\,[(I_4-1)\,\mathbf{I}+(I_1-3)\,\mathbf{M}_1]+K^{\text{coup2}}\,[(I_6-1)\,\mathbf{I}+(I_1-3)\,\mathbf{M}_2]\end{aligned} \tag{1.43}$$

Mandel stress and Second Piola-Kirchoff stress tensors can be further derived by implementing Eqs. (1.38) and (1.37), respectively. The stiffness tensor can be derived by implementing the following equations

$$\mathscr{L} = 4\frac{\partial^2\psi_e}{\partial\mathbf{C}_e^2}. \tag{1.44}$$

The stiffness tenor for the in-plane part of the current model takes the following form

$$\hat{\mathscr{L}}^{ip} = \begin{bmatrix}\hat{\mathscr{L}}^{2\times2}_{\text{stretch}} & 0\\ 0 & \hat{\mathscr{L}}^{1\times1}_{\text{shear}}\end{bmatrix} \tag{1.45}$$

and the final form of the stiffness tensor is

$$\mathscr{L}_{ip} = \left[\begin{array}{c|c|c} 8K_1^{\text{ani1}} & 4K_2^{\text{iso}} + 4K^{\text{coup ani}} & 0 \\ \hline 4K_2^{\text{iso}} + 4K^{\text{coup ani}} & 8K_1^{\text{ani2}} & 0 \\ \hline 0 & 0 & -2K_2^{\text{iso}} \end{array}\right]. \tag{1.46}$$

1.3.3 Plasticity

The yield criterion implemented for the large strains regime is similar to what was used for the small strains regime (1.18)

$$\Phi(\boldsymbol{\Sigma}, R) = \left\{ \sum_{\alpha=1}^{6} \left\{ x_\alpha \left[\frac{\boldsymbol{\Sigma} : \mathbf{H}^\alpha}{r_\alpha^0} \right]^{2k} \right\} \right\}^{\frac{1}{2k}} - (\sigma_0 + R), \tag{1.47}$$

The yield plane normals $\mathbf{H}^\alpha$ can be derived in the same way as in the small strains approach (1.24). The same is applied for r_α^0 as in (1.20). By implementing the associated flow rule, the evolution equation of the plastic flow rate can be expressed by

$$\mathbf{L}_p = \dot{\lambda}\mathbf{N}, \tag{1.48}$$

with $\mathbf{N}$ denoting the direction of the plastic strain rate tensor, defined by

$$\mathbf{N} = \frac{\partial \Phi}{\partial \boldsymbol{\Sigma}} \Bigg/ \left\| \frac{\partial \Phi}{\partial \boldsymbol{\Sigma}} \right\|. \tag{1.49}$$

and the final form of constitutive equations can be summarized below:

$$\begin{aligned} &\boldsymbol{\Sigma} = 2\mathbf{C}_e \frac{\partial \psi_e}{\partial \mathbf{C}_e}, \\ &\mathbf{L}_p = \dot{\lambda}\mathbf{N}, \ \dot{\kappa} = \dot{\lambda}, \\ &\Phi = \left\{ \sum_{\alpha=1}^{6} \left\{ x_\alpha \left[\frac{\boldsymbol{\Sigma} : \mathbf{H}^\alpha}{r_\alpha^0} \right]^{2k} \right\} \right\}^{\frac{1}{2k}} - (\sigma_0 + R), \ R = Q\left(1 - \mathrm{e}^{-\beta\kappa}\right), \\ &\dot{\lambda} \geq 0, \ \Phi \leq 0, \ \dot{\lambda}\Phi = 0. \end{aligned} \tag{1.50}$$

Numerical Implementation

The semi-implicit backward Euler scheme with an exponential map is applied for the solution of the constitutive equations. Based on the definition $\mathbf{L}_p := \dot{\mathbf{F}}_p \mathbf{F}_p^{-1}$, the evolution equation for the flow rate can be written as

$$\dot{\mathbf{F}}_p = \mathbf{L}_p \mathbf{F}_p = \dot{\lambda}\mathbf{N}\mathbf{F}_p. \tag{1.51}$$

Considering a time increment Δt for the time step $[t_n, t_{n+1}]$, the previous equation becomes

$$\mathbf{F}_p = \exp\left(\bar{\lambda}\mathbf{N}^{(0)}\right)\mathbf{F}_{p_n}, \tag{1.52}$$

where $\bar{\lambda}$ is defined as $\bar{\lambda} := \dot{\lambda}\Delta t$ and $\mathbf{N}^{(0)}$ is the plastic flow direction at the beginning of local iterations, and the final form of the equation becomes

$$\begin{aligned} \mathbf{r}_1 &= \mathbf{F}_p - \exp\left(\bar{\lambda}\mathbf{N}^{(0)}\right)\mathbf{F}_{p_n} = \mathbf{0}, \\ r_2 &= \Phi = 0, \end{aligned} \tag{1.53}$$

This set of equations is solved iteratively by implementing the Newton method.

1.4 Model Validation

The process of validating the suggested models involves two main steps; the first step includes the parameters calibration based on the simple tension tests done in MD and CD directions, then the first validation test is done by comparing the model results with the tension tests results done in other directions. The second step includes the comparison of the numerical results with the experimental results for the punch test.

1.4.1 Parameters Calibration

The parameters representing the elastic behavior of the material are first calibrated based on the simple tension tests on MD and CD directions.

1.4.1.1 Small Strains Model

For the small strains model, the necessary parameters are: K_2^{iso}, K_1^{ani1}, K_1^{ani2} and $K^{\text{coup ani}}$. The fitting of the parameters is done only on the elastic part of the experiments. The final values of the elastic parameters can be shown in Table 1.1 The fitting of the parameters describing the behavior of the material in the plastic regime is done by implementing the plastic part of the tension tests on MD and CD directions. The hardening parameters for the small strains model σ_0 and R are fitted

Table 1.1 Elastic parameters fitted from experiments

Parameter	K_1^{ani1}	K_1^{ani2}	$K^{\text{coup ani}}$	K_2^{iso}
Value [MPa]	2966	1034	2212	1313

first, and based on these results, the ratio of initial yield stress in different direction to the reference yield stress R^c_{MD},R^t_{MD},R^c_{CD},R^t_{CD},and $R^t_{45^o}$ can be calculated. Table 1.2 shows the final values for the plastic parameters for the small strains model. To validate the fitting results, the numerical results for the simple tension in MD and CD directions are compared with the experimental results as shown in Fig. 1.3.

1.4.1.2 Large Strains Model

The fitting of the elastic parameters for the large strains model is done in the same way done for small strains model. The values of the calibrated elastic parameters can be found in Table 1.3. In the same way, the plastic parameters for the large strains model can be calibrated. The results are shown in Table 1.4. Based on the calibration results, the numerical results for the large strains model for simple tension test can be shown in Fig. 1.4.

Table 1.2 Plastic parameters for the small strains regime model

Parameter	Definition	Value
k [ ]	Constant	2
d_{MD} [ ]	Ratio of lateral plastic strain to axial plastic strain in MD tension	−0.62 ± 0.05
d_{CD} [ ]	Ratio of lateral plastic strain to axial plastic strain in CD tension	−0.14 ± 0.02
σ_0 [MPa]	Reference initial yield stress	24.8
Q [MPa]	Hardening constant	17.0
β [ ]	Hardening constant	109.3
R^c_{MD} [ ]	Ratio of initial yield stress in MD compression to σ_0	0.43
R^t_{CD} [ ]	Ratio of initial yield stress in CD tension to σ_0	0.42
R^c_{CD} [ ]	Ratio of initial yield stress in CD compression to σ_0	0.28
$R^t_{45^\circ}$ [ ]	Ratio of initial yield stress in 45° tension to σ_0	0.52

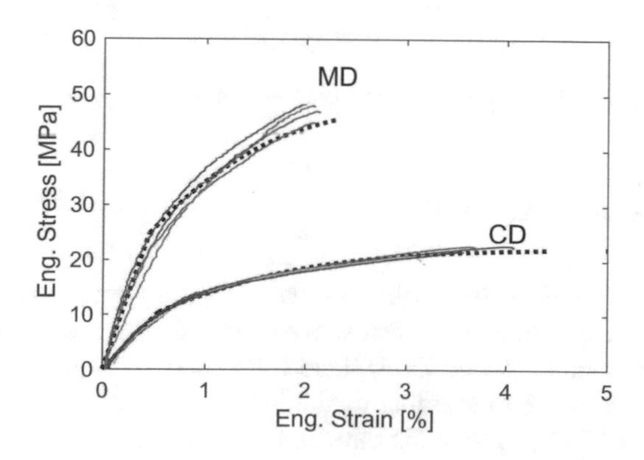

Fig. 1.3 Fitting results for the small strains model

Table 1.3 Elastic parameters fitted from experiments

Parameter	K_1^{ani1}	K_1^{ani2}	$K^{\text{coup ani}}$	K_2^{iso}
Value [MPa]	741	258	1101	-656

Table 1.4 Plastic parameters for the large strains regime model

Parameter	Definition	Value
k [ ]	Constant	2
d_{MD} [ ]	Ratio of lateral plastic strain to axial plastic strain in MD tension	-0.62 ± 0.05
d_{CD} [ ]	Ratio of lateral plastic strain to axial plastic strain in CD tension	-0.14 ± 0.02
σ_0 [MPa]	Reference initial yield stress	15
Q [MPa]	Hardening constant	40
β [ ]	Hardening constant	160
R^c_{MD} [ ]	Ratio of initial yield stress in MD compression to σ_0	0.711
R^t_{CD} [ ]	Ratio of initial yield stress in CD tension to σ_0	0.71
R^c_{CD} [ ]	Ratio of initial yield stress in CD compression to σ_0	0.463
$R^t_{45°}$ [ ]	Ratio of initial yield stress in 45° tension to σ_0	0.86

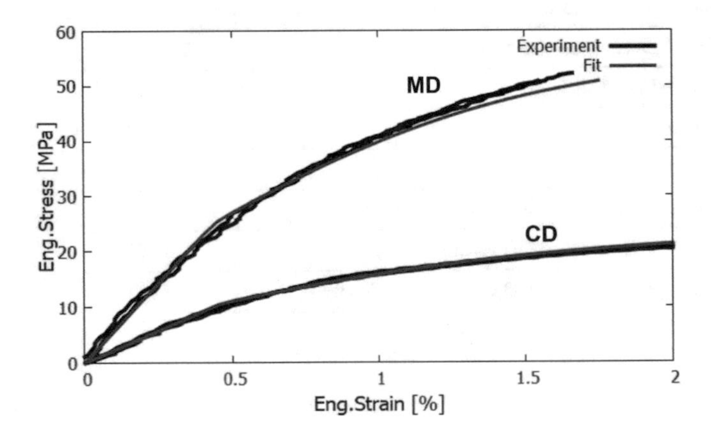

Fig. 1.4 Fitting results for the large strains model

1.4.2 Punch test

Both models are further validated by implementing complex loading conditions. The suggested punch test was performed with the set-up shown in Fig. 1.5. The test included a 'Zwick Z100' load frame and 'Aramis 4M' DIC system, which allowed synchronized recording of the local strain fields. The dimension of the set-up was L = 155 mm, W = 70 mm, and D = 70 mm. Both ends of the specimen were first

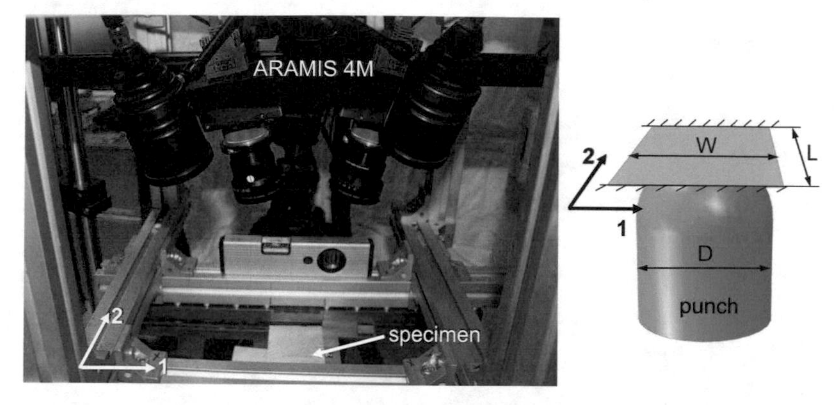

Fig. 1.5 Punch test set-up (1 - MD direction and 2 - CD direction)

clamped to the holders, and the blank holder force was applied to hold the specimen in place through the whole test process. The load was applied to the punch at a crosshead speed of 0.15 mm/s until fracture occurred. The whole punch test was also analyzed using the static implicit solver of Abaqus. Due to the symmetric characteristics of the whole set-up, only one quarter of the specimen was simulated as shown in Fig. 1.6.

The experiment results for force-displacement curve have been compared to the numerical results coming from both of the small and large strains models as shown in Fig. 1.7. The results indicate a good agreement between the simulation and experimental results. The results show also good agreement between both models representing the small and large strains regimes. For further validation, the local strain fields in CD direction are compared at two different displacements (7.92 mm

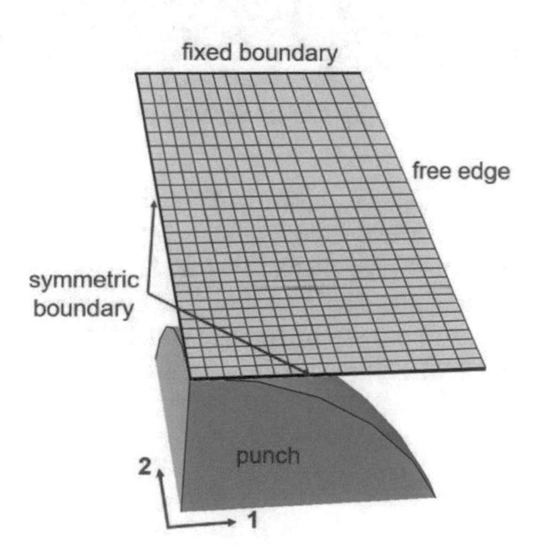

Fig. 1.6 Finite element model of the punch test

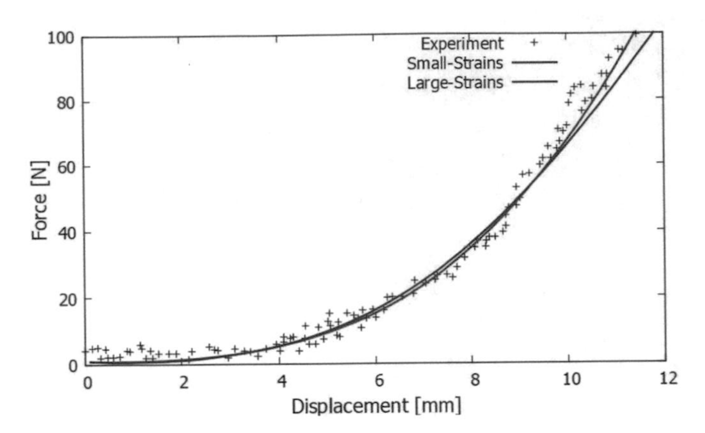

Fig. 1.7 Force-Displacement curves for both experiments and prediction from the small and large strains models

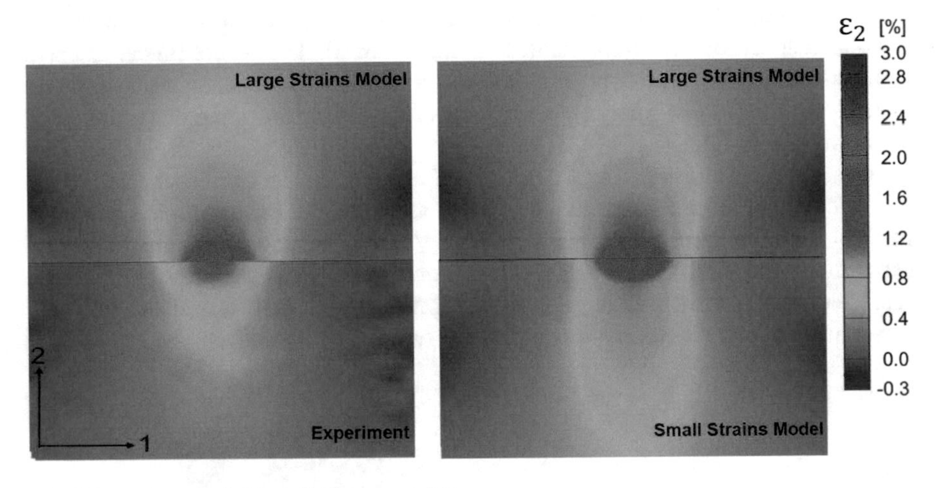

Fig. 1.8 Local strain fields at displacement 7.92 mm

and 13 mm) as shown in Figs. 1.8 and 1.9, respectively. The resulted contour plots indicate that both models can describe the local distribution of strain fields at different displacements. This leads to the conclusion that the extension of the small strains model into large strains can still describe the in-plane behavior correctly and thus can be used for further modelling of paper behavior also in the out-of-plane direction.

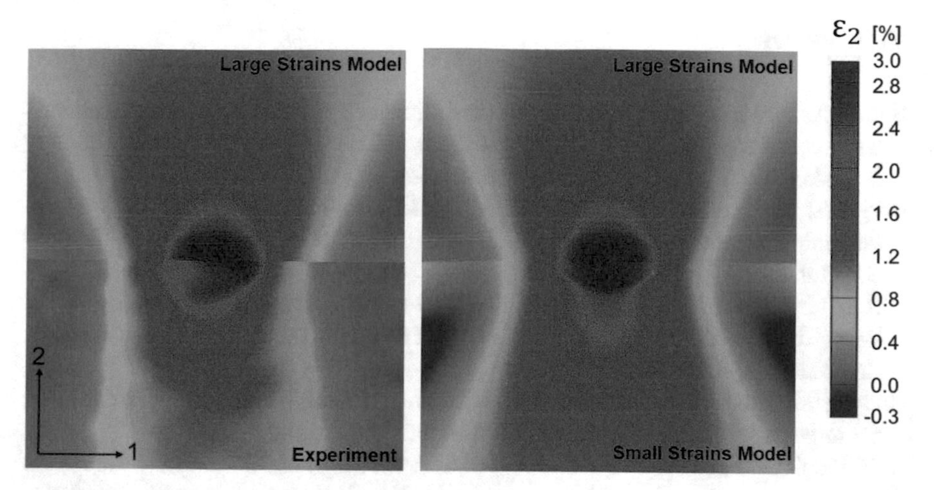

Fig. 1.9 Local strain fields at displacement 13 mm

1.5 Conclusions

The anisotropic behavior of paper in the in-plane direction has been analyzed by implementing two different approaches. The small strains model was constructed by including the concept of the structural tensors to describe the elastic behavior, and the yield function proposed by Li et al (2016) combined with an isotropic hardening law to describe the plastic behavior of the material. The presented model was able to capture the anisotropic behavior of the material which can be observed in simple tension tests. The extension of this model to take into account the large strains was also presented, where the same yield criterion was implemented. This model showed a good agreement with the previous model, which was shown from the punch test results. Both of the small strains and large strains models were able to simulate the anisotropic behavior of the material during punch test. The local strains fields produced by both models were in a reasonable agreement with the experimental results. This yield to the conclusion that the modified model derived in the large strains regime can be adopted for further analysis of paper, and can be extended to account for the large strains observed in thickness direction. Such extension will enable the model to simulate the behavior of paper under complex loading conditions, such as the creasing and folding of paperboard.

Acknowledgements We gratefully acknowledge the scientific support from Prof. Stefanie Reese from RWTH Aachen University.

References

Beex LAA, Peerlings RHJ (2012) On the influence of delamination on laminated paperboard creasing and folding. Philosophical Transactions of the Royal Society a-Mathematical Physical and Engineering Sciences 370(1965):1912–1924

Borgqvist E, Lindstrom T, Tryding J, Wallin M, Ristinmaa M (2014) Distortional hardening plasticity model for paperboard. International Journal of Solids and Structures 51(13):2411–2423

Hill R (1948) A theory of the yielding and plastic flow of anisotropic metals. Proceedings of the Royal Society of London Series A Mathematical and Physical Sciences 193(1033):281–297

Huang H, Hagman A, Nygards M (2014) Quasi static analysis of creasing and folding for three paperboards. Mechanics of Materials 69(1):11–34

Li Y, Stapleton S, Reese S, Simon JW (2016) Anisotropic elastic-plastic deformation of paper: in-plane model. International Journal of Solids and Structures 100-101: 286-296.

Naumenko K, Altenbach H (2016) Modeling High Temperature Materials Behavior for Structural Analysis, Advanced Structured Materials, vol 28. Springer, DOI 10.1007/978-3-319-31629-1

Reese S (2003) Meso-macro modelling of fibre-reinforced rubber-like composites exhibiting large elastic-plastic deformation. International Journal of Solids and Structures 40(4):951 – 980

Reese S, Raible T, Wriggers P (2001) Finite element modelling of orthotropic material behaviour in pneumatic membranes. International Journal of Solids and Structures 38(52):9525–9544

Xia QXS, Boyce MC, Parks DM (2002) A constitutive model for the anisotropic elastic-plastic deformation of paper and paperboard. International Journal of Solids and Structures 39(15):4053–4071

Chapter 2
Experiments on Damage and Fracture Mechanisms in Ductile Metals under Non-proportional Loading Paths

Michael Brünig, Steffen Gerke, and Moritz Zistl

Abstract The paper discusses new experiments to study the effect of non-proportional loading paths on damage and fracture behavior of ductile metals. Specimens are taken from thin metal sheets and are tested under different biaxial loading conditions covering a wide range of stress states. In this context, a thermodynamically consistent anisotropic continuum damage model is presented based on yield and damage conditions as well as evolution laws for plastic and damage strain rates. To validate the constitutive laws and to identify material parameters different experiments with the biaxially loaded X0-specimen have been performed. Results for proportional and corresponding non-proportional loading histories are discussed. During the experiments strain fields in critical regions of the specimens are analyzed by digital image correlation (DIC) technique while the fracture surfaces are examined by scanning electron microscopy (SEM). The results elucidate the effect of loading history on damage and fracture behavior in ductile metals.

Key words: Ductile metals · Biaxial experiments · Damage · Fracture · Non-proportional loading

2.1 Introduction

Demands on reduction in energy consumption as well as improvement in cost efficiency, in safety and in lifetime have led to increased research activities to develop high quality metals during the last decades. Material strengths have to be enhanced to reduce localization of irreversible strains and to avoid damage and fracture of structural elements under different loading conditions and loading histories. Therefore,

Michael Brünig · Steffen Gerke · Moritz Zistl
Institut für Mechanik und Statik, Universität der Bundeswehr München, Werner-Heisenberg-Weg 39, 85577 Neubiberg, Germany,
e-mail: michael.bruenig@unibw.de, steffen.gerke@unibw.de,moritz.zistl@unibw.de

H. Altenbach et al. (eds.), *Plasticity, Damage and Fracture in Advanced Materials*, Advanced Structured Materials 121,
https://doi.org/10.1007/978-3-030-34851-9_2

modeling of material behavior as well as simulation of deformation and failure processes of these new metals are important aspects in engineering applications. Thus, accurately predictive and practically applicable constitutive approaches as well as development of experiments for their validation and identification of material parameters are necessary to fulfill these requests.

It has been observed in several experiments that during loading of material elements localized inelastic strains occur which are accompanied by damage and fracture processes on the micro-level leading to macro-cracks and final failure of structural elements. Characteristics of the damage and fracture processes on the micro-scale depend on the stress state and on the loading histories acting in a material point. Thus, development of suitable continuum models must be based on detailed studies of the stress-state-dependent and loading-path-dependent phenomena. In this context, many experiments with various specimens have been discussed in the literature. For example, dependence of damage and fracture processes on the stress triaxiality has been investigated by uniaxial tests with unnotched and differently notched specimens and corresponding numerical simulations (Bai and Wierzbicki, 2008; Bao and Wierzbicki, 2004; Bonora et al, 2005; Brünig et al, 2008; Driemeier et al, 2010; Dunand and Mohr, 2011; Gao et al, 2010; Roth and Mohr, 2016). Since these uniaxially loaded specimens only cover a small range of stress triaxialities biaxial experiments with different cruciform specimens have been proposed (Brünig et al, 2015a,b, 2018; Green et al, 2004; Kulawinski et al, 2011; Kuwabara, 2007).

In many engineering applications non-proportional loading processes occur and, thus, experimental programs with different non-proportional loading paths have to be developed. For example, tests with non-proportionally loaded notched cruciform specimens have been performed to investigate loading-path-dependent fracture behavior of aluminum alloys (Chow and Lu, 1992; Wang and Chow, 1989). In the tension-tension regime the specimens showed after proportional and non-proportional loading histories remarkable differences in propagation of crack paths and the final fracture loads. Experiments with non-proportional loading paths with stress-triaxiality step jumps showed different fracture loci compared to those with proportional loading (Basu and Benzerga, 2015). In addition, tension-torsion tests with hollow cylindrical steel specimens have been performed with changing the tension to torsion ratio during runs (Cortese et al, 2016). Further tension-torsion and compression-torsion tests with non-proportional loading paths revealed effects of the loading histories on damage accumulation and the fracture initiation loci in the specimens (Zhuang et al, 2016).

Furthermore, micro-mechanical numerical analyses examining deformation behavior of void-containing unit cells have been performed to reveal characteristics of stress-state-dependent damage and fracture processes on the micro-scale, see Brünig et al (2013) and references therein for further details. Micro-mechanical finite element studies with proportional and corresponding non-proportional loading paths showed remarkably different failure loci (Benzerga et al, 2012) and larger increase in damage in material samples undergoing non-proportional loading histories (Roux et al, 2014). Thus, results of the experiments and the micro-mechanical numerical calculations discussed above clearly demonstrate that formation of damage and fracture processes remarkably depend on the loading history and have to be examined in further detail.

Therefore, new biaxial experiments with the recently developed X0-specimen (Gerke et al, 2017) undergoing different non-proportional loading histories with high and moderate positive stress triaxialities have been performed (Gerke et al, 2019). They showed remarkably different damage and fracture mechanisms compared to proportional loading paths and further experimental results with this specimen are discussed in the present paper. To motivate the new experimental program a continuum damage model is discussed with special focus on different stress states and loading histories. In the experiments digital image correlation (DIC) technique is applied to monitor formation of strain fields in critical regions of the specimens where localization of inelastic deformations and first damage and fracture processes are expected to occur. After the tests fracture mechanisms are visualized by scanning electron microscopy (SEM) of the fracture surfaces.

2.2 Continuum Damage Model

Modeling and simulation of inelastic deformations as well as of damage and fracture behavior in ductile metals is based on the continuum damage model proposed by Brünig (2003, 2016). The basic idea of this phenomenological model is the introduction of macroscopic damage strains caused by stress-state-dependent damage and failure mechanisms on the micro-scale. Consideration of damaged and fictitious undamaged configurations is the basis of the kinematic approach using elastic, plastic and damage strain rate tensors. Free energy functions are formulated with respect to the different configurations leading to elastic constitutive equations for the undamaged matrix material and to damage-elastic relations for the damaged material sample taking into account the deteriorating effect of damage on elastic material properties.

Furthermore, in the undamaged configurations the Drucker-Prager-type yield criterion

$$f^{pl} = a\bar{I}_1 + \sqrt{\bar{J}_2} - c = 0 \tag{2.1}$$

predicts onset of plastic behavior of the investigated aluminum alloy AlSiMgMn (EN AW 6082-T6). In Eq. (2.1) the first and second invariants, $\bar{I}_1$ and $\bar{J}_2$, of the effective Kirchhoff stress tensor (Brünig, 2003) with respect to the undamaged configurations are used. The plastic hardening behavior is modeled by the power law for the current equivalent effective stress

$$c = c_0 \left(\frac{H_0\, \gamma}{n\, c_0} + 1 \right)^n \tag{2.2}$$

with the initial yield stress c_0 = 162 MPa, the hardening modulus H_0 = 800 MPa as well as the hardening exponent n = 0.17 and γ is taken to be the equivalent plastic strain measure characterizing the amount of plastic deformations (Brünig, 2003). The isochoric effective plastic strain rate

$$\dot{\bar{\mathbf{H}}}^{pl} = \dot{\gamma}\,\bar{\mathbf{N}} \tag{2.3}$$

models the evolution of plastic deformations where $\bar{\mathbf{N}} = 1/(\sqrt{2\bar{J}_2})\ \mathrm{dev}\bar{\mathbf{T}}$ is the normalized effective deviatoric stress tensor. Validation of the flow rule (2.3) for general applications must be based on different multiaxial experiments undergoing proportional and non-proportional loading paths which are discussed in the present paper.

In addition, in the damaged configurations the damage criterion

$$f^{da} = \alpha I_1 + \beta\sqrt{J_2} - \sigma = 0 \tag{2.4}$$

characterizes onset and continuation of damage based on the damage surface concept formulated in stress space (Brünig, 2003; Chow and Wang, 1987). Assuming isotropic initial elastic and plastic behavior the damage criterion (2.4) is written in terms of the first and second deviatoric stress invariants I_1 and J_2 of the Kirchhoff stress tensor with respect to the damaged configurations. In Eq. (2.4) the damage threshold

$$\sigma = \sigma_0 - H_1\,\mu^2 \tag{2.5}$$

denotes material toughness to micro-defect propagation with the initial equivalent stress $\sigma_0 = 250$ MPa and the modulus $H_1 = 400$ MPa where μ is the equivalent damage strain measure quantifying the amount of irreversible deformations caused by damage. In Eq. (2.4) the parameters α and β represent damage mode variables taking into account different damage mechanisms acting on the micro-level: void-growth-dominated processes for large positive stress triaxialities, shear modes for negative ones and mixed modes (simultaneous growth of voids and evolution of micro-shear-cracks) for moderate positive and nearly zero stress triaxialities. In the proposed continuum damage model, the parameter β also depends on the Lode parameter because it has been shown that its effect on the formation of the micro-structural effects can be remarkable especially in moderate positive and negative stress triaxiality regions (Brünig et al, 2013). Thus, the damage mode parameters α and β in Eq. (2.4) depend on the stress intensity $\sigma_{eq} = \sqrt{3J_2}$ (von Mises equivalent stress), the stress triaxiality

$$\eta = \frac{\sigma_m}{\sigma_{eq}} = \frac{I_1}{3\sqrt{3J_2}} \tag{2.6}$$

defined as the ratio of the mean stress $\sigma_m = I_1/3$ and the von Mises equivalent stress σ_{eq} as well as on the Lode parameter

$$\omega = \frac{2T_2 - T_1 - T_3}{T_1 - T_3} \quad \text{with } T_1 \geq T_2 \geq T_3 \tag{2.7}$$

expressed in terms of the principal values T_1, T_2 and T_3 of the Kirchhoff stress tensor.

The effect of α and β on stress state has been examined in detail for the investigated aluminum alloy. In particular, numerical simulations on the micro-scale have been performed by Brünig et al (2013) analyzing the deformation and failure behavior of void-containing unit cells. With the proposed stress-state-dependent functions the parameters α and β correspond to different damage and fracture mechanisms acting on the micro-level. In addition, based on observations in experiments with biaxially loaded specimens simplified functions for these damage mode parameters have been developed in Brünig et al (2016) for practical applications still allowing accurate simulation of the inelastic deformation as well as of the damage and failure behavior.

Based on these analyses (Brünig et al, 2013; Brünig et al, 2016), the parameter α is taken to be

$$\alpha(\eta) = \begin{cases} -0.15 & \text{for } \eta \leq 0 \\ 0.33 & \text{for } \eta > 0 \end{cases} \tag{2.8}$$

whereas the parameter β is given by the non-negative function

$$\beta(\eta, \omega) = -1.28\,\eta + 0.85 - 0.017\,\omega^3 - 0.065\,\omega^2 - 0.078\,\omega \ \geq \ 0\ . \tag{2.9}$$

Moreover, the damage rule leads to the damage strain rate tensor

$$\dot{\mathbf{H}}^{da} = \dot{\mu}\left(\bar{\alpha}\,\frac{1}{\sqrt{3}}\,\mathbf{1} + \frac{\bar{\beta}}{\sqrt{2}}\,\mathbf{N}\right) \tag{2.10}$$

where $\dot{\mu}$ is a non-negative scalar-valued factor. In Eq. (2.10) the stress related deviatoric tensor $\mathbf{N} = 1/(\sqrt{2J_2})\ \mathrm{dev}\tilde{\mathbf{T}}$ has been used. The stress-state-dependent parameters $\bar{\alpha}$ and $\bar{\beta}$ are kinematic variables denoting the portion of volumetric and isochoric damage-induced strain rates. These variables also correspond to various damage and fracture mechanisms on the micro-level and, similar to the parameters in the damage criterion (2.4), they have been developed performing numerical analyses with micro-defect-containing representative volume elements undergoing different three-dimensional loading scenarios (Brünig et al, 2013) as well as by comparison of experimental and numerical results of various tests with uniaxially and biaxially loaded specimens (Brünig et al, 2016).

In particular, the stress-state-dependence of the variable $\bar{\alpha}$ corresponding to the amount of increase in volumetric damage-induced deformations caused by volume changes of micro-defects is given by

$$\bar{\alpha}(\eta) = \begin{cases} 0 & \text{for } \eta \leq 0 \\ 0.5714\,\eta & \text{for } 0 < \eta \leq 1.75 \\ 1 & \text{for } \eta > 1.75 \end{cases}\ . \tag{2.11}$$

In addition, the stress-state-dependence of the parameter $\bar{\beta}$ corresponding to the amount of increase in anisotropic isochoric damage strains caused by evolution of micro-shear-cracks is expressed in the form

$$\bar{\beta}(\eta, \omega) = \bar{\beta}_0(\eta) + (-0.0252 + 0.0378\,\eta)\,\bar{\beta}_\omega(\omega) \tag{2.12}$$

with

$$\bar{\beta}_0(\eta) = \begin{cases} 0.87 \ \text{ for } \ \eta \leq \frac{1}{3} \\ 0.979 - 0.326\,\eta \ \text{ for } \ \frac{1}{3} < \eta \leq 3 \\ 0 \ \text{ for } \ \eta > 3 \end{cases} \tag{2.13}$$

and

$$\bar{\beta}_\omega(\omega) = \begin{cases} 1 - \omega^2 \ \text{ for } \ \eta \leq \frac{2}{3} \\ 0 \ \text{ for } \ \eta > \frac{2}{3} \end{cases} \ . \tag{2.14}$$

It can be clearly seen that in the macroscopic damage rule (2.10) a volumetric part (first term in Eq. (2.10)) corresponding to isotropic growth of voids on the micro-level as well as a deviatoric part (second term in Eq. (2.10)) associated with anisotropic development of micro-shear-cracks, respectively, are taken into account. Thus, the basic damage processes discussed above (isotropic growth of voids and formation of micro-shear-cracks) acting on the micro-scale are involved in the macroscopic damage rule (2.10) in a phenomenological way. It is worthy to note that since the damage rule (2.10) can be used in different engineering applications its validation must also be based on a number of multiaxial experiments with proportional and non-proportional loading paths which are discussed for an aluminum alloy in the present paper.

2.3 Biaxial Experiments with the X0-specimen

A biaxial experimental program has been presented by Gerke et al (2017) with new specimens analyzing the effect of the stress state on inelastic deformations as well as on damage and fracture processes in ductile metals. In addition, within biaxial tests different loading histories can be taken into account with proportional and non-proportional loading paths (Gerke et al, 2019) and results of further test with the X0-specimen are discussed in the present paper. The experiments are performed with the biaxial test machine shown in Fig. 2.1(a). It contains four electro-mechanically driven cylinders with loads up to ±20 kN located in perpendicular axes and the biaxial specimens are fixed in the heads of the cylinders with clamped boundary conditions. Furthermore, during the experiments three-dimensional displacement fields in selected parts of the specimens are monitored by digital image correlation (DIC) technique. The stereo setting contains four 6.0 Mpx cameras with corresponding lighting system shown in Fig. 2.1(b).

The investigated material is the aluminum alloy AlSiMgMn (EN AW 6082-T6) and specimens are taken from thin sheets with 4 mm thickness. The X0-specimen shown in Fig. 2.2(a) is characterized by four crosswise arranged bars with a central opening and four notched regions inclined by 45° where localization of inelastic deformations, damage and fracture is expected to occur (Fig. 2.2(d)). The specimens dimensions are 240 mm in each axis and the depth of the notches is 1 mm reducing the thickness in these parts from 4 mm to 2 mm at the thinnest points (Fig. 2.2(c)). These notched parts have the length of 6 mm (Fig. 2.2(b)) whereas the notch radii are 3 mm in plane and 2 mm in thickness directions, respectively. During the tests

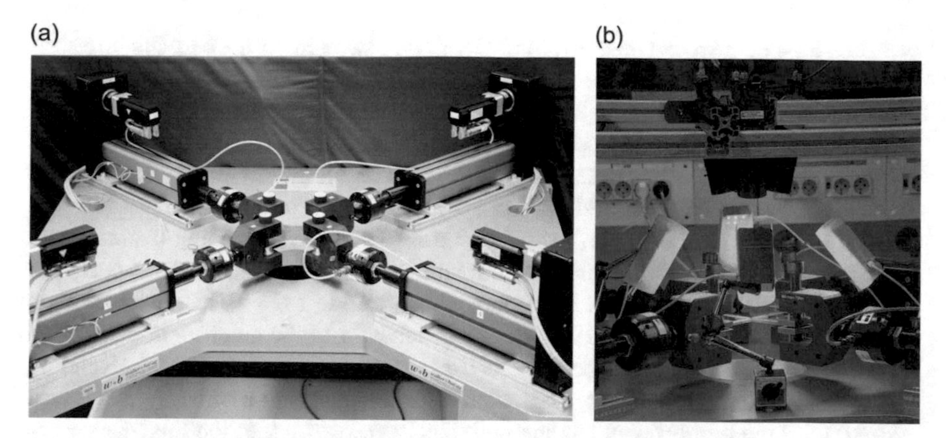

Fig. 2.1 (a) Biaxial test machine, (b) lighting system and camera equipment

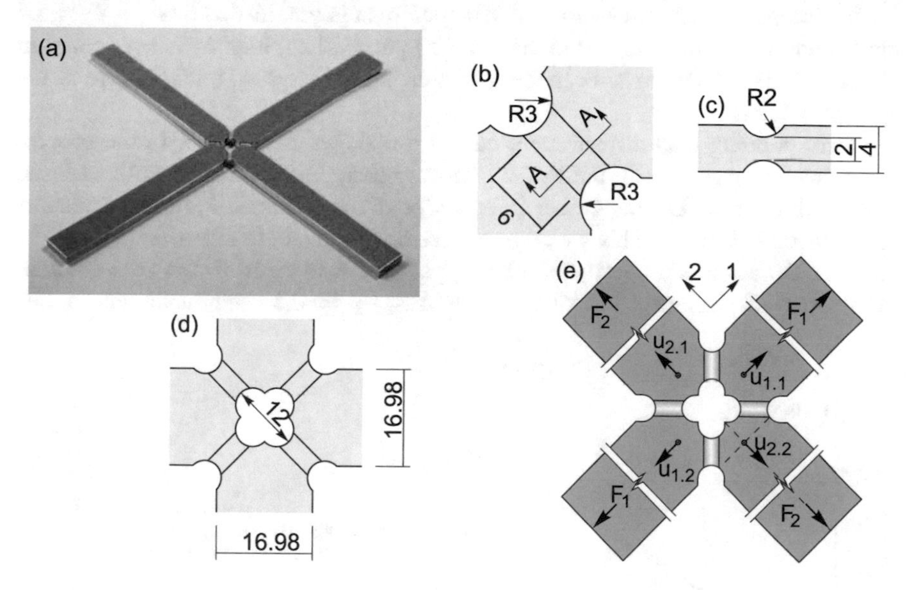

Fig. 2.2 X0-specimen (all dimensions in mm): (a) photo, (b) notch geometry in specimen plane, (c) notch geometry in thickness direction, (d) specimen's center, (e) forces and displacements

the displacements of the red points, $u_{1.1}$ and $u_{1.2}$ in 1-direction and $u_{2.1}$ and $u_{2.2}$ in 2-direction, shown in Fig. 2.2(e) are recorded by DIC. This leads to the relative movements $\Delta u_{ref.1} = u_{1.1} - u_{1.2}$ and $\Delta u_{ref.2} = u_{2.1} - u_{2.2}$, respectively, used in the load-displacement curves. In addition, the displacements normal to the plane are controlled to reveal possible buckling during compressive loading. In all experiments discussed in the present paper these normal displacements were marginal and no buckling took place.

The X0-specimen can be individually loaded in two directions with the forces F_1 and F_2 (Fig. 2.2(e)) and the investigated loading conditions are shown in Fig. 2.3 for proportional and two different non-proportional loading paths. Biaxial loading of the specimen leads to tensile and shear deformations as well as to combination of these basic processes. These deformations cause different damage and fracture mechanisms on the micro-level with positive, nearly zero and negative stress triaxialities. In the present paper, focus is on non-proportional loading histories and the obtained experimental results are compared with those taken from proportional loading paths. In particular, in the proportional loading case (P 1/0) the specimen is only loaded by F_1 leading to tension-shear behavior in the notches. In the second case (NP 1/-1 to 1/0) the X0-specimen is first loaded by $F_1 = -F_2 = 3.9$ kN leading to shear behavior in the notches. Then, load F_1 is kept constant whereas F_2 is reduced to zero until the proportional loading path has been reached. In the final step, F_1 is increased until fracture of the specimen occurs. In the alternative non-proportional loading scenario (NP 1/+1 to 1/0) the X0-specimen is first loaded by $F_1 = F_2 = 5.4$ kN leading to tensile behavior in the notched parts. Then, load F_1 is kept constant whereas F_2 is reduced to zero. In the final step, F_1 is increased until fracture of the specimen occurs.

Corresponding load-displacement curves are shown in Fig. 2.4. In the proportional loading path (P 1/0) the load F_1 shows nearly linear increase until inelastic deformations occur. During further deformation of the specimen F_1 further increases up to fracture at $F_1 = 6.5$ kN with the relative displacement in axis 1 (a1) $\Delta u_{ref.1} =$ 1.0 mm. During this proportional loading path the load F_2 is always zero whereas the relative displacement in axis 2 (a2) is $\Delta u_{ref.2}$ = -0.57 mm at the end of the

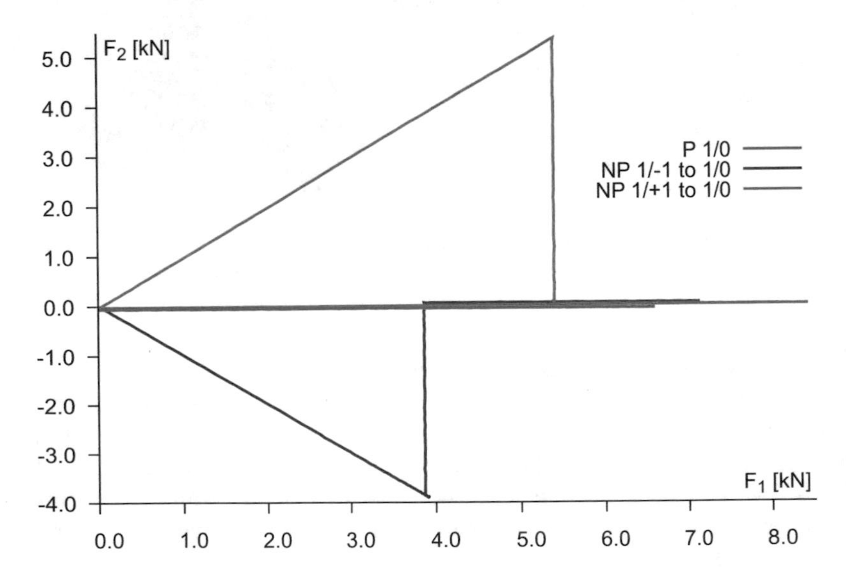

Fig. 2.3 Loading paths

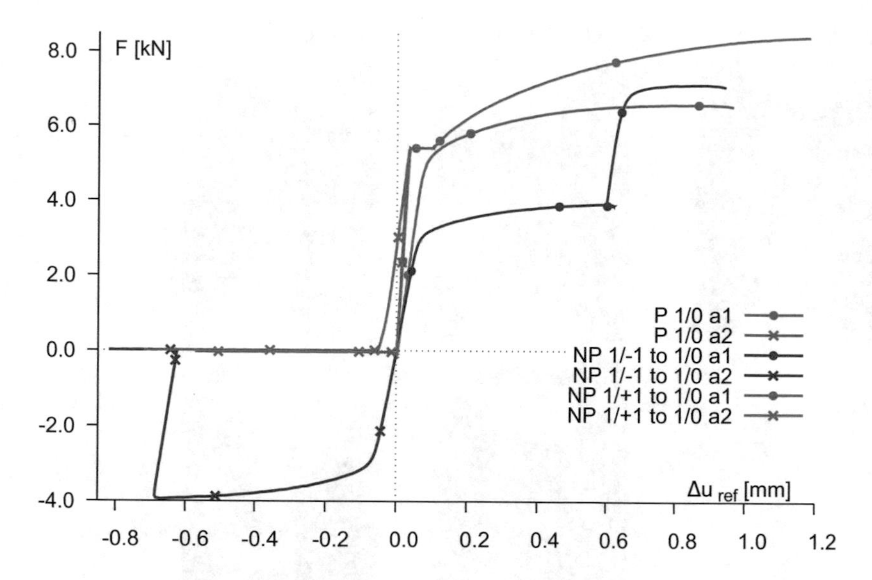

Fig. 2.4 Load-displacement curves

test. In the first non-proportional case (NP 1/-1 to 1/0) the load F_1 increases linearly in the elastic part and non-linearly in the subsequent inelastic one up to $F_1 = 3.9$ kN and the corresponding displacement in axis 1 is $\Delta u_{ref.1} = 0.61$ mm. Similar behavior can be seen in axis 2 with compressive loading up to $F_2 = -3.9$ kN and the displacement $\Delta u_{ref.2} = -0.69$ mm. In the next step, load F_2 is reduced to zero and after this unloading path the displacement in axis 2 is $\Delta u_{ref.2} = -0.62$ mm whereas the load F_1 is kept constant and the corresponding displacement in axis 1 reduces to $\Delta u_{ref.1} = 0.60$ mm. In the final proportional step F_1 further increases first linearly indicating elastic behavior with subsequent non-linear curve characteristic for inelastic behavior. The final load at fracture is $F_1 = 7.1$ kN and the corresponding displacement in axis 1 is $\Delta u_{ref.1} = 0.94$ mm. In axis 2 the final displacement is $\Delta u_{ref.2} = -0.82$ mm. In the alternative non-proportional loading scenario (NP 1/+1 to 1/0) both loads F_1 and F_2 increase up to $F_1 = F_2 = 5.4$ kN and the corresponding displacements are $\Delta u_{ref.1} = \Delta u_{ref.2} = 0.10$ mm. In the next step, load F_2 is reduced to zero and after this unloading the displacement in axis 2 is $\Delta u_{ref.2} =$ -0.05 mm whereas the load F_1 is kept constant and the corresponding displacement in axis 1 only marginally changes. In the final proportional step F_1 further increases non-linearly indicating inelastic behavior. The final load at fracture is $F_1 = 8.4$ kN and the corresponding displacement in axis 1 is $\Delta u_{ref.1} = 1.19$ mm. In axis 2 the final displacement $\Delta u_{ref.2} = -0.72$ mm is reached. These load-displacement curves clearly show that the loading history remarkably affects the final loads at fracture as well as the corresponding displacements.

During biaxial testing of the X0-specimen displacement fields have been monitored by DIC and the distributions of the first principal strains are shown in Fig. 2.5.

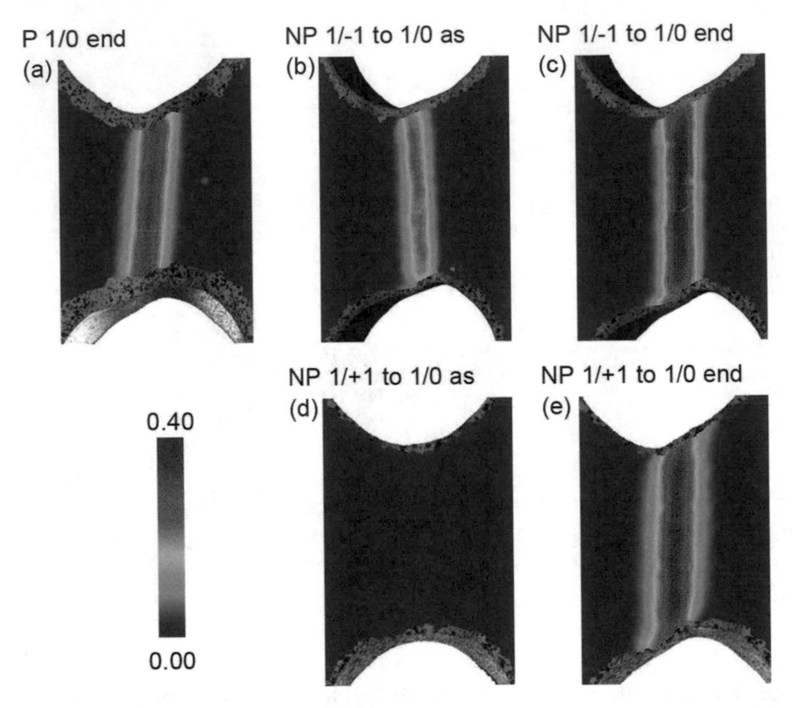

Fig. 2.5 First principal strain fields

In particular, at the end of the proportional loading path (P 1/0 end) a localized band of strains can be seen in the notched part of the specimen with maximum values of 34% with the orientation from right to left (Fig. 2.5(a)). In the first non-proportional loading case (NP 1/-1 to 1/0) after the first loading scenario with $F_1 = -F_2$ before axis switch (as) a localized band of strains occurs with maximum values of 25% and an orientation from left to right (Fig. 2.5(b)). After further loading with final step on the proportional path (end) again a localized band of strains can be seen with remarkable principal strains of 40%. During this loading step, the orientation of the band changes and finally it is nearly vertical (Fig. 2.5(c)). In the alternative non-proportional loading path (NP 1/+1 to 1/0) after the first loading scenario with $F_1 = F_2$ before axis switch (as) the strains remain small with less than 10% (Fig. 5(d)). After further loading with final step on the proportional path (end) again a localized band of strains can be seen with principal strains of 37% (Fig. 2.5(e)) but the band is less localized compared to those of the other loading histories (see Figs. 2.5(a) and (c)). The orientation of the band is again from right to left but less pronounced compared to the proportional case (Fig. 2.5(a)). Thus, the loading history has an influence on the width and orientation of the localized bands of the first principal strain as well as on its amount.

Photos of the central parts of the fractured specimens are shown in Fig. 2.6. The fracture lines correspond to the location and orientation of the bands of maximum principal strains and only show slight differences. In addition, Fig. 2.6 also shows

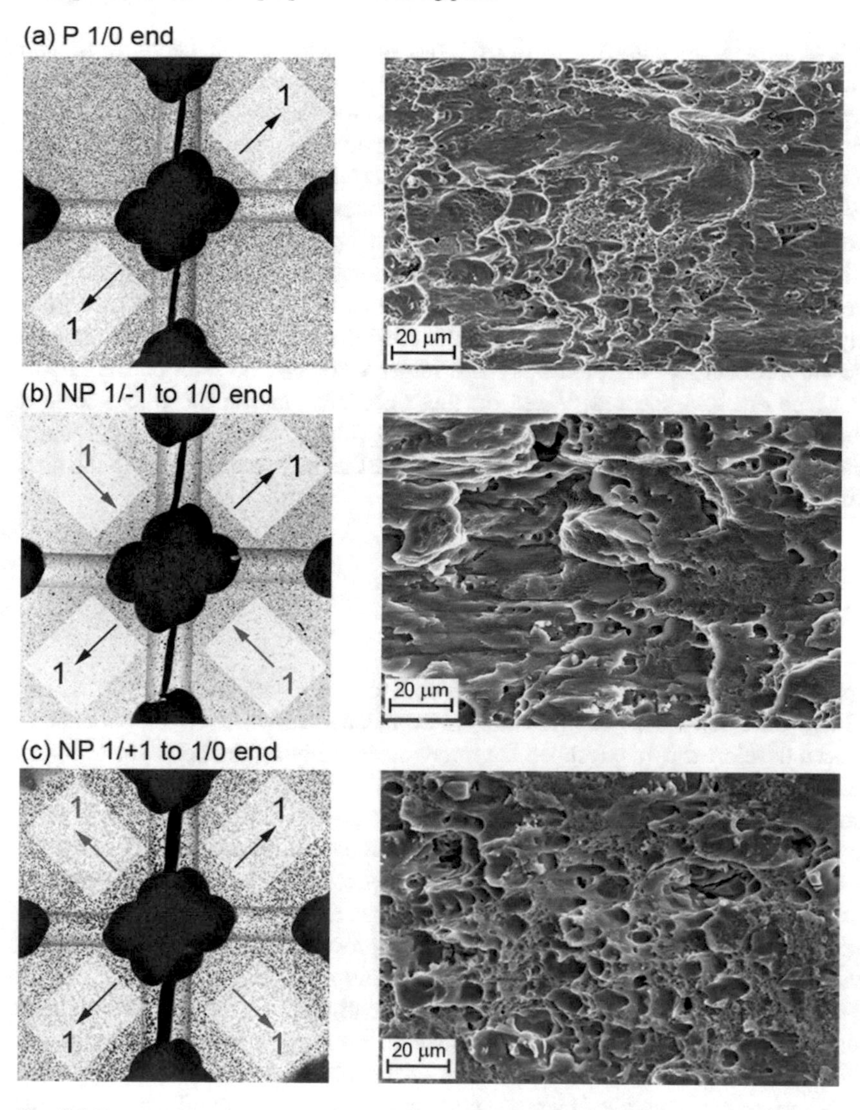

Fig. 2.6 Fractured specimens and SEM pictures of the fracture surfaces

SEM pictures of the fracture surfaces. In particular, at the end of the proportional loading path (P 1/0 end) growth of voids in combination with shear effects can be seen which is the typical damage and fracture mechanism for the tension-shear deformation behavior in the notched regions of the X0-specimen under the only load F_1. On the other hand, at the end of the non-proportional loading path (NP 1/-1 to 1/0 end) remarkable shear effects in combination with few voids are visualized by SEM. First loading with $F_1 = -F_2$ causes shear deformation behavior in the notches leading on the micro-scale to micro-shear-cracks. After unloading of F_2 additional final loading with F_1 causes growth of some voids with simultaneous formation of further micro-shear-cracks leading to the shear-predominated fracture behavior shown in Fig. 2.6(b). Compared to the proportional loading history the shear effects are more predominant with less and smaller voids on the micro-level. Furthermore, after the alternative non-proportional loading path (NP 1/+1 to 1/0 end) remarkable voids can be seen in Fig. 2.6(c) which are slightly sheared and superimposed by few micro-shear-cracks. In this case first loading with $F_1 = F_2$ causes tensile stresses with high portion of hydrostatic stress due to the notches leading on the micro-level to predominant growth of voids. After decrease of F_2 to zero further loading in axis 1 only leads to further growth of voids with superimposed shear effects. Compared to the proportional loading path more and larger voids can be seen on the micro-scale which are less sheared. Based on these observations on the micro-scale it can be concluded that the loading path remarkably affects the damage and fracture mechanisms and the processes occurring firstly are the predominant ones in the final fracture process.

2.4 Conclusions

The paper has discussed a continuum framework to model damage of ductile materials. Stress-state-dependent functions for micro-mechanically motivated parameters have been developed by analysis on the micro-scale studying the behavior of three-dimensionally loaded void-containing representative volume elements. To validate the continuum damage model and the proposed stress-sate-dependent functions new experiments with the biaxially loaded X0-specimen have been performed and results have been compared with those taken from corresponding numerical simulations. Focus was on different loading paths with the same final loading ratio. The experimental investigations revealed the effect of non-proportional loading paths on the damage and fracture behavior in ductile metals compared to proportional ones. To motivate the investigations a phenomenological continuum damage model has been discussed taking into account different branches in the damage criteria corresponding to different stress-state-dependent processes on the micro-scale. Evolution of plastic and damage strains is modeled by rate equations which have to be validated by series of biaxial experiments undergoing proportional and corresponding non-proportional loading histories. Therefore, further results of biaxial tests with the X0-specimen have been presented. In the critical notched regions of the specimen various shear-tension behaviors are caused by different proportional and non-proportional biaxial loading paths leading to different strain states as well as to different stress-state-dependent damage and fracture processes on the micro-level. Thus, evolution of damage and fracture mechanisms is remarkably affected by the loading history and have to be considered in validation of accurate material models predicting failure and lifetime of engineering structures.

Acknowledgements The project has been funded by the Deutsche Forschungsgemeinshaft (DFG, German Research Foundation) – project number 322157331, this financial support is gratefully acknowledged. The SEM images of the fracture surfaces presented in this paper were performed at the Institut für Werkstoffe im Bauwesen, Bundeswehr University Munich and the support of Wolfgang Saur is gratefully acknowledged.

References

Bai Y, Wierzbicki T (2008) A new model of metal plasticity and fracture with pressure and Lode dependence. International Journal of Plasticity 24(6):1071 – 1096

Bao Y, Wierzbicki T (2004) On fracture locus in the equivalent strain and stress triaxiality space. International Journal of Mechanical Sciences 46(1):81 – 98

Basu S, Benzerga A (2015) On the path-dependence of the fracture locus in ductile materials - experiments. International Journal of Plasticity 71:79–90

Benzerga AA, Surovik D, Keralavarma SM (2012) On the path-dependence of the fracture locus in ductile materials – analysis. International Journal of Plasticity 37:157 – 170

Bonora N, Gentile D, Pirondi A, Newaz G (2005) Ductile damage evolution under triaxial state of stress: theory and experiments. International Journal of Plasticity 21(5):981 – 1007

Brünig M, Gerke S, Schmidt M (2016) Biaxial experiments and phenomenological modeling of stress-state-dependent ductile damage and fracture. International Journal of Fracture 200:63–76

Brünig M (2003) An anisotropic ductile damage model based on irreversible thermodynamics. International Journal of Plasticity 19(10):1679 – 1713

Brünig M (2016) A thermodynamically consistent continuum damage model taking into account the ideas of CL Chow. International Journal of Damage Mechanics 25(8):1130–1141

Brünig M, Chyra O, Albrecht D, Driemeier L, Alves M (2008) A ductile damage criterion at various stress triaxialities. International Journal of Plasticity 24(10):1731 – 1755, special Issue in Honor of Jean-Louis Chaboche

Brünig M, Gerke S, Hagenbrock V (2013) Micro-mechanical studies on the effect of the stress triaxiality and the Lode parameter on ductile damage. International Journal of Plasticity 50:49 – 65

Brünig M, Brenner D, Gerke S (2015a) Modeling of stress-state-dependent damage and failure of ductile metals. Applied Mechanics and Materials 784:35–42

Brünig M, Brenner D, Gerke S (2015b) Stress state dependence of ductile damage and fracture behavior: Experiments and numerical simulations. Engineering Fracture Mechanics 141:152 – 169

Brünig M, Gerke S, Schmidt M (2018) Damage and failure at negative stress triaxialities: Experiments, modeling and numerical simulations. International Journal of Plasticity 102:70 – 82

Chow CL, Lu TJ (1992) An analytical and experimental study of mixed-mode ductile fracture under nonproportional loading. International Journal of Damage Mechanics 1(2):191–236

Chow CL, Wang J (1987) An anisotropic theory of continuum damage mechanics for ductile fracture. Engineering Fracture Mechanics 27(5):547 – 558

Cortese L, Nalli F, Rossi M (2016) A nonlinear model for ductile damage accumulation under multiaxial non-proportional loading conditions. International Journal of Plasticity 85:77–92

Driemeier L, Brünig M, Micheli G, Alves M (2010) Experiments on stress-triaxiality dependence of material behavior of aluminum alloys. Mechanics of Materials 42(2):207 – 217

Dunand M, Mohr D (2011) On the predictive capabilities of the shear modified Gurson and the modified Mohr-Coulomb fracture models over a wide range of stress triaxialities and Lode angles. Journal of the Mechanics and Physics of Solids 59:1374–1394

Gao X, Zhang G, Roe C (2010) A study on the efect of the stress state on ductile fracture. International Journal of Damage Mechanics 19:75–94

Gerke S, Adulyasak P, Brünig M (2017) New biaxially loaded specimens for the analysis of damage and fracture in sheet metals. International Journal of Solids and Structures 110-111:209 – 218

Gerke S, Zistl M, Bhardwaj A, Brünig M (2019) Experiments with the X0-specimen on the effect of non-proportional loading paths on damage and fracture mechanisms in aluminum alloys. International Journal of Solids and Structures 163:157 – 169

Green D, Neale K, MacEven S, Makinde A, Perrin R (2004) Experimental investigation of the biaxial behaviour of an aluminum sheet. International Journal of Plasticity 20:1677–1706

Kulawinski D, Nagel K, Henkel S, Hübner P, Fischer H, Kuna M, Biermann H (2011) Characterization of stress-strain behavior of a cast trip steel under different biaxial planar load ratios. Engineering Fracture Mechanics 78:1684–1695

Kuwabara T (2007) Advances in experiments on metal sheet and tubes in support of constitutive modeling and forming simulations. International Journal of Plasticity 23:385–419

Roth C, Mohr D (2016) Ductile fracture experiments with local proportional loading histories. International Journal of Plasticity 79:328–354

Roux E, Shakoor M, Bernacki M, Bouchard P (2014) A new finite element approach for modeling ductile damage void nucleation and growth - analysis of loading path effect on damage mechanisms. Modeling and Simulation in Materials Science and Engineering 22:1–23

Wang J, Chow CL (1989) Mixed mode ductile fracture studies with nonproportional loading based on continuum damage mechanics. Trans ASME Journal of Engineering Materials Technology 111(2):204–209

Zhuang X, Wang T, Zhu X, Zhao Z (2016) Calibration and application of ductile fracture criterion under non-proportional loading conditions. Engineering Fracture Mechanics 165:39–56

Chapter 3
Strength Differential Effect in Martensitic Stainless Steel Under Quenching and Partitioning Heat Treatment Condition

Sebastian Dieck, Martin Ecke, Paul Rosemann, Sebastian Fritsch, Martin Franz-Xaver Wagner, and Thorsten Halle

Abstract The quenching and partitioning (Q&P) heat treatment enables a higher formability of high strength martensitic steels. Therefore it is necessary to have some metastable austenite in the microstructure, which transforms in martensite during plastic deformation (transformation induced plasticity - TRIP effect). This microstructure can be achieved by the two-step heat treatment, consisting of quenching in a way, which retains a certain amount of austenite. A subsequent low temperature annealing, the so called partitioning, stabilizes the retained austenite due to carbon diffusion. The Q&P heat treatment was investigated for the martensitic stainless steel 1.4034 (X46Cr13, AISI 420) concerning the influence of varying austenite fractions. In line with these efforts the characterization of the mechanical properties, focusing on the materials behavior under different mechanical load scenarios, were performed. Therefore quasi static tension and compression tests were carried out. Moreover, a comprehensive analysis of the microstructural evolution was performed for different stages of heat treatment, including optical microscopy and electron backscatter diffraction. The comparison of common quenching and tempering with the Q&P heat treatment verifies the extensively enhanced materials strength with good ductility whereat the formability is still acceptable. The microstructural explanation was an higher austenite fraction due to austenite retaining between the martensite laths besides the stabilizing of retained austenite. Further a strength differential effect was observed to be much higher than known for tempered martensite. Our investigations show first results and cannot clearly demonstrate the complex

Sebastian Dieck · Martin Ecke · Paul Rosemann · Thorsten Halle
Otto von Guericke University Magdeburg, Institute of Materials and Joining Technology, Universitätsplatz 2, 39106 Magdeburg, Germany,
e-mail: sebastian.dieck@ovgu.de, martin.ecke@ovgu.de, paul.rosemann@ovgu.de, thorsten.halle@ovgu.de

Sebastian Fritsch · Martin Franz-Xaver Wagner
Technical University Chemnitz, Faculty of Mechanical Engineering, Professorship Materials Science, Erfenschlager Straße 73, 09125 Chemnitz, Germany,
e-mail: sebastian.fritsch@mb.tu-chemnitz.de, martin.wagner@mb.tu-chemnitz.de

H. Altenbach et al. (eds.), *Plasticity, Damage and Fracture in Advanced Materials*, Advanced Structured Materials 121,
https://doi.org/10.1007/978-3-030-34851-9_3

microstructural mechanism. Furthermore we find some interesting differences in the micromechanical behavior compared to the literature with regard to micro-cracking and no TRIP effect under compressive loading.

Key words: Quenching · Partitioning · Reversion · Martensite · Austenite · Strength differential effect

3.1 Introduction

The current trends in light weight product development represents a challenge for conventional materials. There is a need for new low cost materials with the a special combination of properties. The heat treatment concept of "Quenching and Partitioning" (Q&P) enables the production of high strength martensitic steel with enhanced formability, which originates from a special martensitic austenitic dual phase microstructure (Wang and Speer, 2013). The Q&P heat treatment consists of two main steps as Fig. 3.1 shows. In the first step the steel is austenitised and quenched. The quenching is interrupted at a temperature between martensite start (M_s) and finish temperature (M_f) to produce a dual phase microstructure of martensite and retained austenite. The subsequent Partitioning is a low temperature tempering and leading to the redistribution of Carbon (C) by diffusion. Carbon diffuses from the supersaturated martensite to retained austenite (Edmonds et al, 2006; Speer et al, 2005; Santofima et al, 2011). Thereby, the austenite is stabilized to room temperature enabling the transformation of martensite by mechanical loading (transformation induced plasticity - TRIP effect) and the tetragonal distortion of the martensitic lattice decreases. Both, the softening of the martensite and the stabilization of retained austenite, leads to an increased formability, without a loss of strength (Zhang et al, 2014).

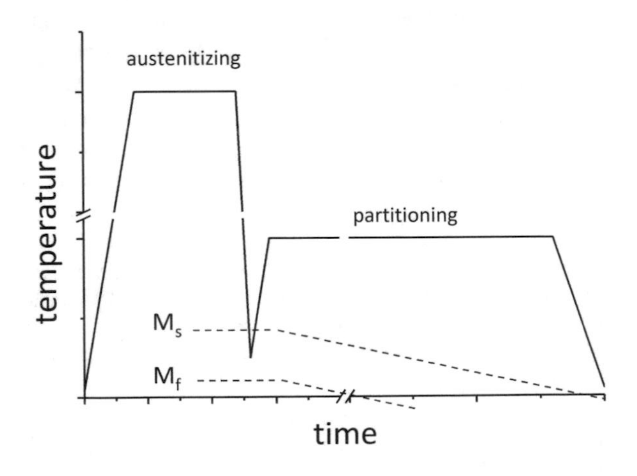

Fig. 3.1 Schematic illustration of Q&P heat treatment process (after Tsuchiyama et al, 2012)

Typical materials used for Q&P heat treatment are low alloyed steels, containing less than 0,3 % C by weight due to weldability efforts. Furthermore, up to 3 wt.% Mn are added to increase the austenite stability and hardenability, as well as up to 2 wt.% Si suppress cementite formation (Wang and Speer, 2013). In contrast the present study focuses on the Q&P treatment of martensitic stainless steel. Previous studies have shown that the Q&P treatment of martensitic stainless steel not only enhances the corrosion resistance (Lu et al, 2016) but also leads to well balanced mechanical properties (Tsuchiyama et al, 2012; Isfahany et al, 2011). Especially the grade AISI 420 is an outstanding candidate for Q&P treatment leading to a tensile strength of 2.000 MPa combined with an elongation of up to 20 % (Yuan et al, 2012). The comparison of the mechanical behavior of a Q&P treated AISI 420 was different between tension and compression loading, attributed to a distinctive strength differential effect (SD effect) (Dieck et al, 2017).

The SD effect is known as the difference of an initial tensile and compression strength and its loading direction depended flow stresses (Ellermann and Scholtes, 2015; Sing et al, 2000; Rauch and Leslie, 2000). For tempered martensite the strength differences are between 5 and 10 % (Ellermann and Scholtes, 2015). There are some possible reasons for the SD effect discussed in literature. The first is about non-linear stresses originating from the production (Bauschinger effect), which slabbing the tensile strength (Hirth and Cohen, 1970). A further theory attributed the SD effect to micro-cracking in brittle structures, e.g. martensite, lead to accelerated fraction under tensile loading, whereas under compression these cracks not tend to open. Additionally, an influence of residual stresses and the circumstance that under compression loading there is no austenite transformation (TRIP) are supposed (Ellermann and Scholtes, 2015; Sing et al, 2000; Rauch and Leslie, 2000; Hirth and Cohen, 1970).

3.2 Experimental

The used material was Fe - 0,44 C - 13,02 Cr - 0,31 Ni - 0,32 Mn - 0,31 Si stainless steel. The heat treatment included austenitising at 1150 °C for 15 min, followed by water quenching with a cooling rate of about 400 K/s. The subsequent Partitioning was performed at 400 °C for 30 min, finished by water quenching. For metallographic analysis the samples were grinded and polished. In the case of optical microscopy the samples were contrasted by etching. Optical microscopy was performed by using ZEISS Axioplan 2 with AxioCam MRc. Electron microscopy was done with FEI Scios with EBSD (AMETEK-EDAX, Hikari Kamera, TSL-OIM 7). The mechanical behavior was examined by using Zwick/Roell UPM 1475. The initial strain rate was 10^{-3} s^{-1} at room temperature. Tensile testing was examined with flat test pieces (type E, corresponding to DIN 50125) and cylindrical pieces for compression testing (10 x 6 mm, corresponding to DIN 50106). The abstraction of samples for mechanical test was parallel to the rolling direction of the material.

3.3 Results and Discussion

First the initial microstructure after Q&P heat treatment was analyzed, Fig. 3.2. The microstructure consists of tempered martensite with a large amount of embedded austenite (3.2 a, b). Furthermore, nm sized carbide precipitations can be found inside of the martensite laths (3.2 c). The line-like austenite distribution in c) points to carbide cluster, remaining from the raw material [10].

The mechanical behavior is shown in Fig. 3.3. An initial flow stress (at 0,2 % of plastic strain) of 1.080 MPa and an ultimate tensile strength (UTS) of 1.725 MPa were determined. These values are slightly lower than expected from the literature. The reason can be identified by SEM analysis of the fracture surface (Fig. 3.3, b). The terraced like cracking in addition with the austenite distribution (Fig. 3.2 c)) suggest the existence of line like carbide distribution, remaining from hot rolling and soft annealing of the raw material. These carbide lines are preferred cracking paths. That's why there is crack splitting and a high amount of brittle fracture, resulting in a reduced formability under tensile loading.

In contrast there is no material failing under compression loading, Fig. 3.4. It have to be observed, that the sample was loaded up to the limits of the testing machine without cracking. The flow stress was 1.400 MPa. Because of the premature end of the test the compression strength could not be determined. For comparison with tensile strength an alternative strength had to be determined. Therefore, the loading

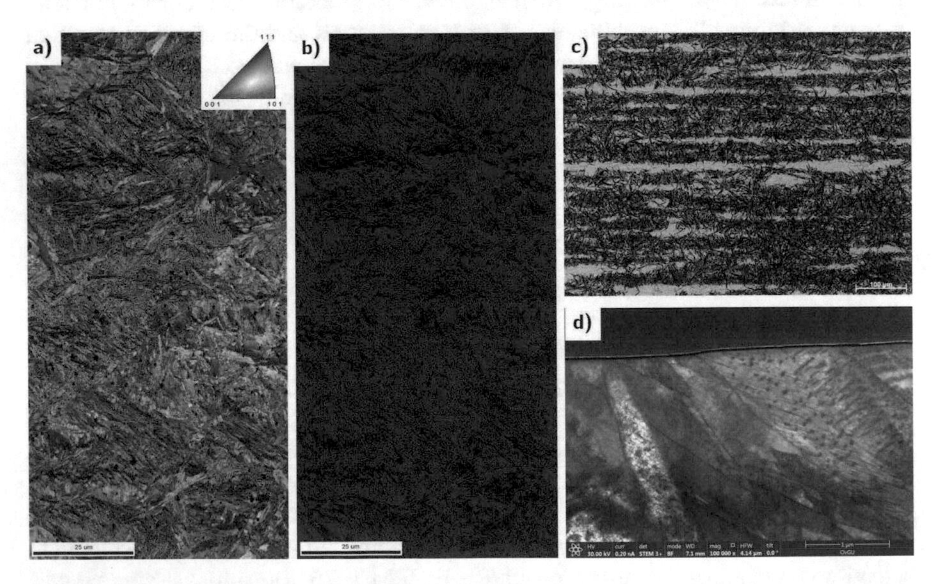

Fig. 3.2 AISI420 after Q&P heat treatment - microstructure characterization by EBSD (a) IPF, b) phase map: martensite: red - 60 %, austenite: blue - 40 %), optical microscopy (c) and STEM (d): microstructure consists of tempered martensite with embedded austenite and nm sized carbide precipitations, the line-like austenite distribution in points to carbide cluster

at 10 % of deformation (tension: $R_{t,10}$= 1.680 MPa; compression: $R_{c,10}$= 2.185 MPa) was taken for comparison of maximum strength.

Comparing both loading scenarios shows, that there is a distinctive strength difference regarding flow stresses (1.080 to 1.725 MPa - SD = 30 %, emanating from the tensile values) as well as comparable maximum stresses (1.680 to 2.185 MPa - SD = 30 %, emanating from the tensile values). This strength differences are much higher than known from literature for tempered martensite (from 5 to 10 %). [11] The reasons for the SD effect in general are not clarified at all. The fact of the distinctive differences makes the investigation much more challenging.

The existence of an internal Bauschinger effect as a consequence of phase transformation during the heat treatment can not be completely excluded but is supposed to be neglected. Otherwise a notable directed plastic pre-strain would effectuate a characteristic change in the stress-strain behavior, which could not be detected for the investigated material conditions. The theory of micro-cracks in brittle

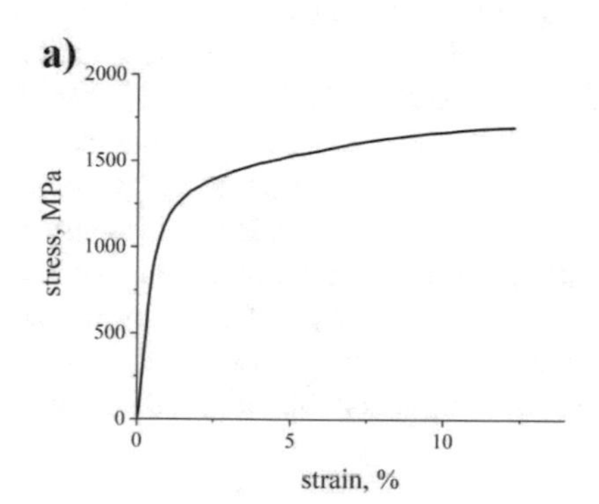

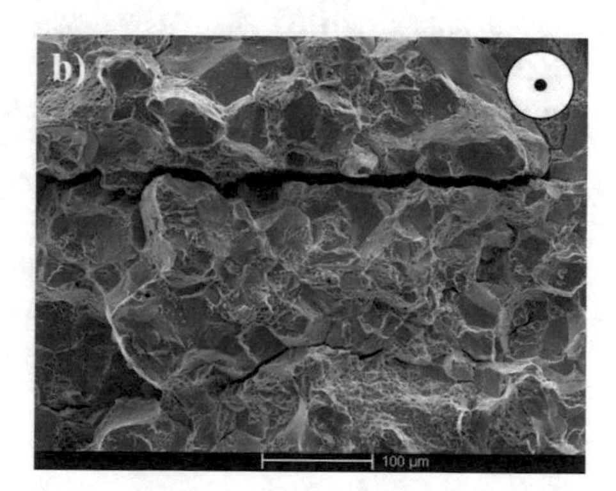

Fig. 3.3 Representative mechanical behavior of AISI420 after Q&P heat treatment for tensile loading (a) and a SEM picture of the fracture surface of the tensile sample (b, loading direction pictured in the upper right corner)

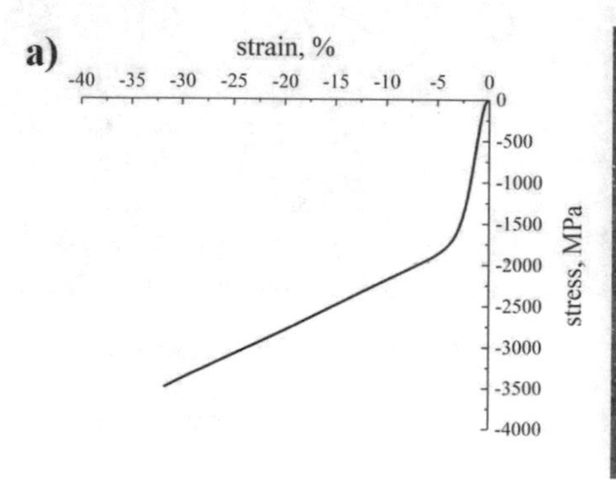

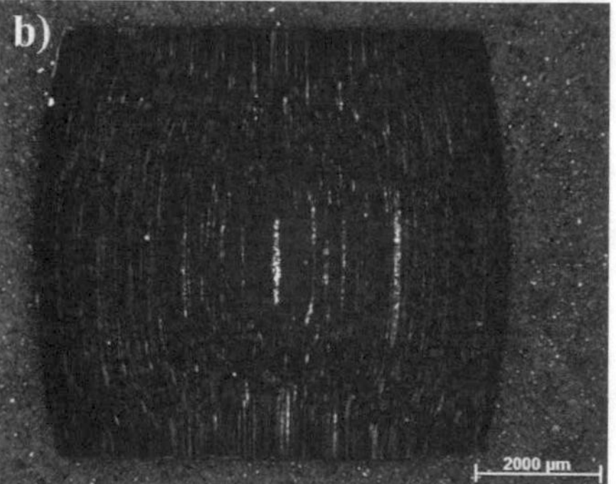

Fig. 3.4 Representative mechanical behavior of AISI420 after Q&P heat treatment for compression loading (a) and cross section of the compression sample (b)

structures, which tend to open only under tensile loading, could be disproved by the comprehensive microscopic investigation. There were no micro-cracks detected for all material states, which confirms the study by Rauch and Leslie. [13] However, the existence of large remaining carbides leads to macro-cracking and limits the maximum elongation of the material. The assumption of macro residual stresses as the reason for the SD effect can not be confirmed. If residual compressive stress, induced by martensitic hardening, remains after the partitioning, an enhanced tensile behavior would be expected.

The hypothesis, that the transformation of retained austenite in martensite is possible for the tensile loading condition but not for compression, could be disproved for Q&P treated martensitic stainless steel. As can be seen from the EBSD phase map comparison (Fig. 3.5), there is a phase transformation for the tensile loading conditions as well as for the compression loading. It can be seen that for the initial state (Fig. 3.2) the austenite fraction is significantly higher (initial: 40 %, tension: 19 %, compression: 8,5 %). This is in contrast to other literature findings (Ellermann and Scholtes, 2015). The TRIP effect of austenite is strongly influenced by its chemical composition and its morphology (Weiß et al, 2009). In contrast to the investigation of (Ellermann and Scholtes, 2015), who investigated the 42CrMoS4 and 100Cr6, the present study observed the behavior of stabilized austenite due to partitioning, which might be the reason for the differing results. Emanating from the assumption that the deformation of the Q&P treated material under loading ends, when all the austenite is transformed in martensite, it can be concluded that the materials limits were not achieved during the mechanical tests of this study. The further existence of austenite in the cross sections of the loaded samples is caused by carbide induced premature cracking of the tensile sample and the enforced stopping of the test for the compression sample.

Fig. 3.5 EBSD phase maps (martensite: red, austenite: blue, CI < 0,1: black - excluded from measuring) for comparing the microstructures of the tensile sample after fracture (left, austenite fraction: 19 %) and the compression sample after maximum loading (right, austenite fraction: 8,5 %)

3.4 Conclusion

Within the present study the mechanical behavior of the AISI 420 in Q&P heat treatment condition was investigated. The maximum elongation under tensile loading is strongly influenced by the conditions of raw material production. Certain conditions during material processing can promote carbide precipitation and agglomeration, which are preferred areas of crack propagation, resulting in prematurely failure. The comparison of mechanical properties for tensile and compression loading proves the existence of an distinctive SD effect of 30 % - much higher than previously measured maximum for tempered martensite (10 %), see Ellermann and Scholtes (2015). For reasons the BAUSCHINGER effect as well as micro-cracking and macro residual stresses could be excluded. In opposite to the literature (Ellermann and Scholtes, 2015) it was found, that under compression loading strain induced transformation of austenite in martensite can be observed as well as under tensile loading.

Acknowledgements The authors would like to acknowledge financial support by the GKMM 1554 and the MDZWP e.V. Furthermore we like to thank the German Research Foundation and federal state Saxony-Anhalt for large equipment founding.

References

Dieck S, Ecke M, Rosemann P, Halle T (2017) Reversed austenite for enhancing ductility of martensitic stainless steel. In: Stebner AP, Olson GB (eds) Proceedings of the International Conference on Martensitic Transformations (Chicago), Springer, Cham, The Minerals, Metals & Materials Series, vol 181, p 012034, DOI 10.1007/978-3-319-76968-4$_$19

Edmonds DV, He K, Rizzo FC, Cooman BCD, Matlock DK, Speer JG (2006) Quenching and partitioning martensite—a novel steel heat treatment. Materials Science and Engineering: A 438-440:25 – 34, DOI 10.1016/j.msea.2006.02.133, proceedings of the International Conference on Martensitic Transformations

Ellermann A, Scholtes B (2015) The strength differential effect in different heat treatment conditions of the steels 42crmos4 and 100cr6. Materials Science and Engineering: A 620:262 – 272, DOI 10.1016/j.msea.2014.10.027

Hirth JP, Cohen M (1970) On the strength-differential phenomenon in hardened steel. Metallurgical Transactions 1(1):3–8, DOI 10.1007/BF02819235

Isfahany AN, Saghafian H, Borhani G (2011) The effect of heat treatment on mechanical properties and corrosion behavior of AISI420 martensitic stainless steel. Journal of Alloys and Compounds 509(9):3931–3936, DOI 10.1016/j.jallcom.2010.12.174

Lu SY, Yao KF, Chen YB, Wang MH, Chen N, Ge XY (2016) Effect of quenching and partitioning on the microstructure evolution and electrochemical properties of a martensitic stainless steel. Corrosion Science 103:95 – 104, DOI 10.1016/j.corsci.2015.11.010

Rauch GC, Leslie WC (2000) The extent and nature of the strength-differential effect in steels. Metallurgical and Materials Transactions B 3(2):377 – 385, DOI 10.1007/BF02642041

Santofima MJ, Zhao L, Sietsma J (2011) Overview of mechanisms Involved During the Quenching and Partitioning Process in Steels. Metallurgical and Materials Transactions A 42:3620–3627, DOI 10.1007/s11661-011-0706-z

Sing AP, Padmanabhan KA, Pandey GN, Murty GMD, Jha S (2000) Strength differential effect in four commercial steels6. Journal of Materials Science 35:1379 – 1388, DOI 10.1023/A:1004738326505

Speer JG, Assunção FCR, Matlock DK, Edmonds DV (2005) The "quenching and partitioning" process: background and recent progress. Materials Research 8:417 – 423, DOI 10.1590/S1516-14392005000400010

Tsuchiyama T, Tobata J, Tao T, Nakada N, Takaki S (2012) Quenching and partitioning treatment of a low-carbon martensitic stainless steel. Materials Science and Engineering: A 532:585 – 592, DOI 10.1016/j.msea.2011.10.125

Wang L, Speer JG (2013) Quenching and partitioning steel heat treatment. Metallography, Microstructure, and Analysis 2(4):268–281, DOI 10.1007/s13632-013-0082-8

Weiß A, Gutte H, Jahn A, Scheller P (2009) Nichtrostende stähle mit trip/twip/sbip-effekt. Materialwissenschaft und Werkstofftechnik 40(8):606–611, DOI 10.1002/mawe.200800361

Yuan L, Ponge D, Wittig J, Choi P, Jiménez JA, Raabe D (2012) Nanoscale austenite reversion through partitioning, segregation and kinetic freezing: Example of a ductile 2gpa fe–cr–c steel. Acta Materialia 60(6):2790 – 2804, DOI 10.1016/j.actamat.2012.01.045

Zhang K, Liu P, Li W, Guo Z, Rong Y (2014) Ultrahigh strength-ductility steel treated by a novel quenching–partitioning–tempering process. Materials Science and Engineering: A 619:205 – 211, DOI 10.1016/j.msea.2014.09.100

Chapter 4
Deformation Twinning in bcc Iron - Experimental Investigation of Twin Formation Assisted by Molecular Dynamics Simulation

Martin Ecke, Oliver Michael, Markus Wilke, Sebastian Hütter, Manja Krüger, and Thorsten Halle

Abstract This work deals with investigations on twin formation in bcc pure iron during high strain rate impact experiments. The temperature and grain size effects on the twin formation are discussed with respect to experimental results. The study presents opportunities and limits in experimental twinning observation. In addition to experimental limitations, an introduction in molecular dynamics simulation as suitable tool in twinning investigation is given. For that reason, molecular dynamics simulation is used to visualize twin formation and describe the nucleation mechanism. The results show a very good correlation to the experimental observations.

Key words: Twinning · bcc iron · High strain rate deformation · Electron backscatter diffraction · Transmission electron microscopy · Molecular dynamics simulation

4.1 Introduction

Deformation twinning is an important mechanism of plastic deformation in metals. There are many material applications and technical issues where deformation twinning can contribute to plastic deformation. Examples are rolling of hcp (hexagonal closed packed) magnesium alloys (Kadiri and Oppedal, 2010) or the TWIP-effect used in fcc (face centered cubic) high manganese alloyed steels (Krüger et al, 2011;

Martin Ecke · Oliver Michael · Markus Wilke · Sebastian Hütter · Thorsten Halle
Otto von Guericke University Magdeburg, Institute of Materials and Joining Technology, Universitätsplatz 2, 39106 Magdeburg, Germany,
e-mail: martin.ecke@ovgu.de, oliver.michael@st.ovgu.de, markus.wilke@ovgu.de, sebastian.huetter@ovgu.de, thorsten.halle@ovgu.de

Manja Krüger
Forschungszentrum Jülich GmbH, Institute of Energy and Climate Research (IEK), Microstructure and Properties of Materials (IEK-2), 52425 Jülich, Germany,
e-mail: manja.krueger@ovgu.de

H. Altenbach et al. (eds.), *Plasticity, Damage and Fracture in Advanced Materials*, Advanced Structured Materials 121,
https://doi.org/10.1007/978-3-030-34851-9_4

Gutierrez-Urrutia et al, 2010). Even in bcc (body centered cubic) iron meteorites, deformation twins were observed and investigated (Neumann, 1849; Ecke et al, 2014). In the past century, understanding deformation twinning was the aim in many scientific works. There are some frequently cited papers giving a comprehensive overview about the formation of twins in bcc materials with respect to several boundary conditions like crystal structure, temperature, grain size or strain rate. Publications by Christian and Mahajan (1995); Meyers et al (2001) are assuredly one of the most cited.

Besides the twin formation depending on the mentioned parameters, the mechanism behind twin nucleation and twin growth are an equally important research focus. For bcc iron, several mechanisms requiring dislocation reactions are discussed. Beginning with the pole mechanism by Cottrell and Bilby (1951), who predicted a dislocation dissociation of a 1/2 [111] perfect dislocation as nucleation reaction. A later extension of this theory was delivered by Sleeswyk (1963). In result, a generated 1/6 [111] dislocation may cross-slip onto another plane functioning as nucleation point of a twin. Another discussed mechanism is the breakaway of a single partial dislocation predicated by Suzuki and Barrett (1958) and Priestner and Leslie (1965). Based on a pole reaction, a stacking-fault can be generated. First experimental evidence of three-layer micro twins were observed in Mo-Re and Fe-Cr-Co by (Mahajan, 2002). There are further mechanisms, based on dislocation reactions and even dislocation-free mechanisms like the $\alpha - \varepsilon - \alpha$ phase transformation (Bancroft et al, 1956; Wang et al, 2013).

Nevertheless, an insight in dislocation processes requires a high experimental effort or is impossible to realize. Therefore, this work presents experimental results on the dependence of twinning in bcc iron due to temperature and grain size. Based on experimental limits, molecular dynamics simulation is provided as a tool to observe twin nucleation and twin growth depending on several crystallographic features.

4.2 Experimental Procedures

There are many influences on deformation twinning particular in bcc iron. Presented here is the formation of twins in pure iron after impact experiments at varying temperature and grain size. A hardened steel bolt with an impact velocity of 50 m/s was shot on a sample with 6 mm in diameter and height. In this experimental set-up (Fig. 4.1), the accelerated bolt hits the sample once. After the impact, the accelerated sample is caught in a bullet trap. Using a special sample holder, the plastic deformation of the sample due to the impact is reduced to less than 1 % plastic strain. The experiments are done at room temperature and at -196° C by using liquid nitrogen. A heat treatment at 1150° C and different times was used to obtain grain sizes between 40 μm and 600 μm.

To observe the twin formation depending on temperature and grain size, several microstructural analysis are available. In this study, investigations by means of optical microscopy (OM) and scanning electron microscopy (SEM) combined with electron

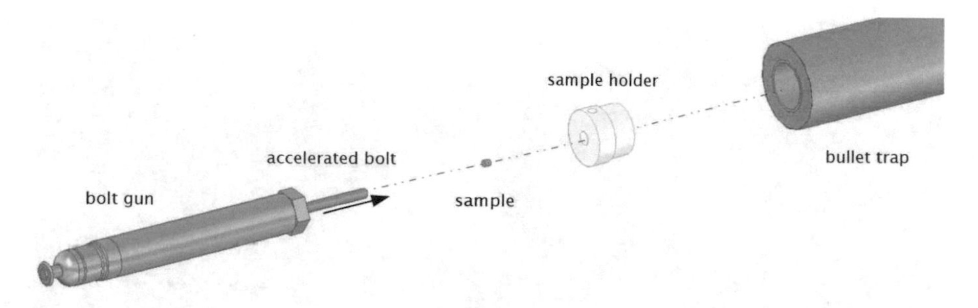

Fig. 4.1 Illustration of experimental set-up; the accelerated bolt hits the sample with an impact velocity of 50 m/s

backscatter diffraction (EBSD) are used. Based on the investigation results, the amount of twins is determined by image processing. For OM results, the analysis of twin formation based on different gray values and shapes and for EBSD results, on different crystallographic properties.

A correlation between the twin amount and the experimental varied temperatures and grain sizes shows a favored twin formation with increasing grain size and decreasing temperature. For example, Fig. 4.2 shows representative OM results for pure iron after impact experiments. The temperature varied between room temperature (a) and -196° C (b). The grain size has a value of 200 μm in average for both microstructures. Figure 4.3 shows EBSD results for impact loaded pure iron at -196° C for 40 μm and 200 μm.

The results shown above deal with the influence of temperature and grain size on the twin formation and the amount of twins. A reduced dislocation agility at low temperatures and decreased dislocation densities and critical shear stresses can be an explanation for the observed behavior. The microstructural observations show the suitability of OM and EBSD to investigate the influences of temperature and grain

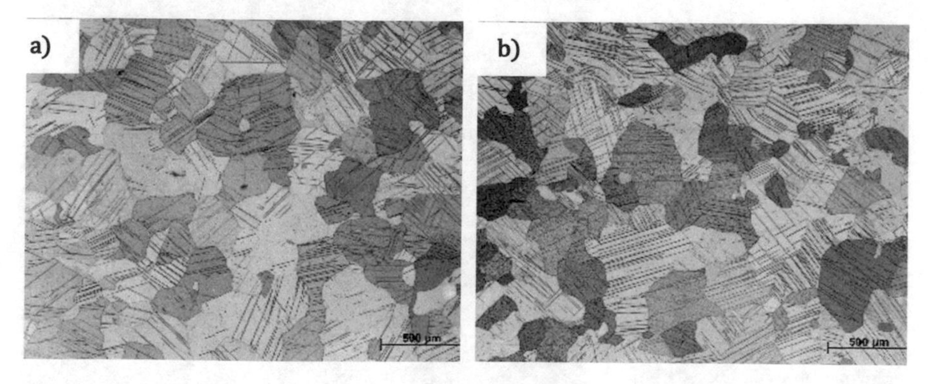

Fig. 4.2 Optical microscopy results of pure iron with 200 μm grain size in average after impact experiments at different temperatures: a) room temperature and b) -196° C

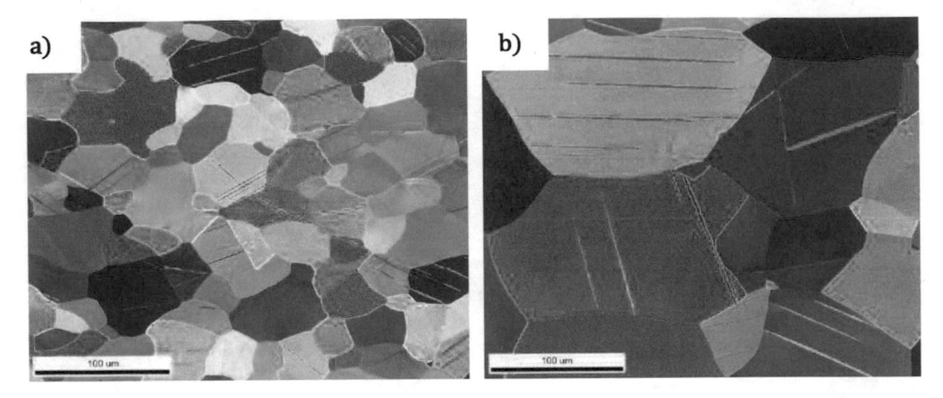

Fig. 4.3 IPF color-coded orientation distribution based on EBSD analysis of pure iron after impact experiments at -196° C for different grain sizes: a) 50 μm and b) 200 μm

size on the twin formation. However, understanding the process of twin nucleation and twin growth is important as well. To investigate the nucleation mechanism in bcc iron and to validate the mechanisms described in literature, transmission electron microcopy (TEM) provide a suitable method. For that reason, a TEM study investigated twin interfaces in pure iron after impact experiments. The TEM samples are lamellae with less than 100 nm thickness prepared by means of focused ion beam technique (FIB). In this first study, dislocation structures obtained on the twin interface, see Fig. 4.4. This result indicates a dislocation-based twinning reaction. Nevertheless, a conclusive interpretation is not possible in this state of investigations.

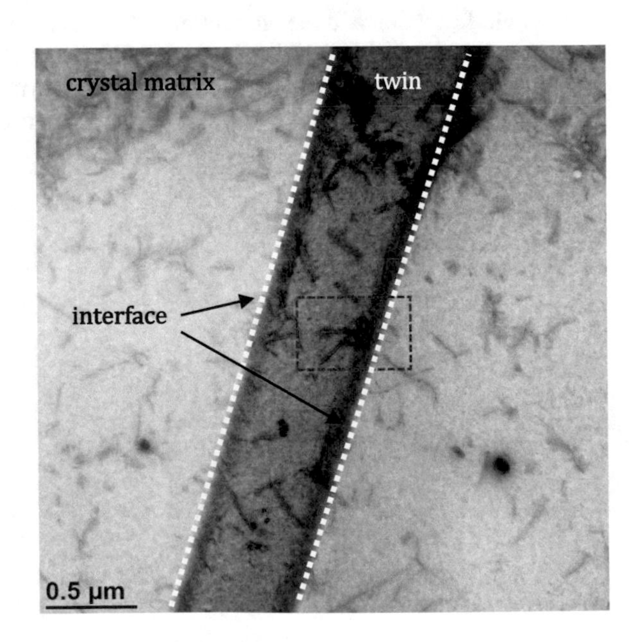

Fig. 4.4 TEM-bright field image of pure iron after impact loading showing a deformation twin. Dislocation structures on the twin interface are observable (marked in red)

Besides the high effort for preparation and analyses, measurement artifacts, probably caused by nanoprecipitations and local stress, complicates the investigation. Further experiments with modified parameters are required to understand the nucleation more in detail.

4.3 Molecular Dynamics Simulation

As mentioned above, observing the twin nucleation experimentally is impossible due to the complex experimental setup that would be needed. To investigate atomic scale processes of impact experiments, molecular dynamics simulation might be a helpful tool. Molecular dynamic is a numerical method at atomic scale to describe interactions between individual atoms. The technique is based on the summation of forces between single atoms derived from a local potential function. These forces can then be integrated using Newtonian laws of motion to obtain local displacements, thus giving full atomic trajectories. Due to the computation time required, only a small volume consisting of a few million atoms can be realistically examined (Zepeda-Ruiz et al, 2017). For the atomistic simulations in this study, simulations are performed using the LAMMPS software (Plimpton, 1995) with a Finnis-Sinclair potential for pure iron (Mendelev et al, 2003). Data post-processing and visualization is done using OVITO (Stukowski, 2009).

To obtain realistic results, a simulation domain is prepared to resemble the volume around a grain boundary where twinning was observed in experiments. To achieve this, the crystal orientation obtained from experimental EBSD results is used to prepare an interface area of two grains. The simulation volume to exhibit twin formation is chosen as 230 x 230 x 143 Å (Fig. 4.5). Time integration is performed by a NVE ensemble integrator with an added thermostat to fix the system temperature at 30 K. To simulate the impact loading, the lower boundary of the simulation box was moved with a constant velocity of 150 m/s and then allowed to relax.

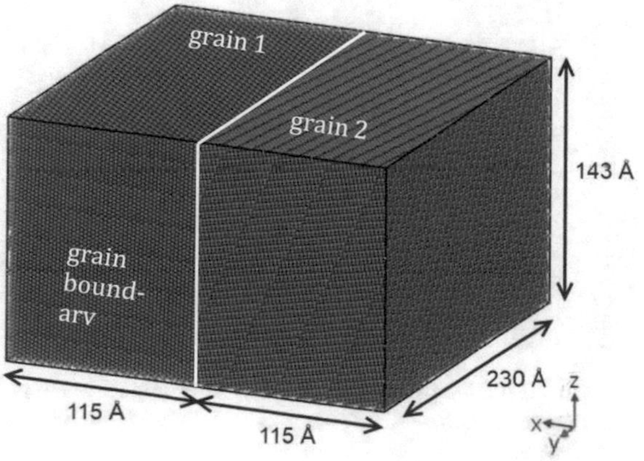

Fig. 4.5 Simulation setup used in molecular dynamics simulation. Periodic boundary conditions are prescribed in left-right and front-back directions, top surface is fixed, bottom surface move with constant velocity of 150 m/s

As molecular dynamics only consider atoms, investigation relating to dislocation reactions requires an additional post-processing step. Using the dislocation extraction algorithm (Stukowski and Albe, 2010; Stukowski et al, 2012), which is implemented in OVITO as well as available as a separate program, lattice types, planar defects, dislocation lines and their Burgers vector can be automatically identified from trajectory files. This method is used to automatically identify twins and examine the dislocation reactions involved in their formations.

A cut through the sample is shown in Fig. 4.6, using color-coded mapping equivalent to that used in EBSD analysis. The formation of two twins with orientations corresponding to the experimental results can be observed. Further investigation of the visualized trajectories shows that twins always nucleate near crystal defects such as grain boundaries and free surfaces, supporting the theory that reactions involving defects and not only shearing of full crystal planes are relevant for the formation of twins.

Further post-processing of these results, using the dislocation extraction algorithm of OVITO, allows direct observation of the dislocation types and reactions involved in the formation and growth of these twins (Fig. 4.7). The analysis shows a number of different dislocation types. Leading the reaction ahead of the tip of the newly formed twin is a 1/2 [111] perfect dislocation, the only type that should be energetically stable in a bcc crystal. Therefore, the presence of some crystal defects is necessary for the formation of twins. Concentration of shear stresses at some point on this dislocation allows the dissociation of this dislocation into two unstable partial dislocation of types 1/6 [111] and 1/3 [111], of which the former form the boundary of the new twin.

Looking back at the various theories reviewed in the introduction, this reaction is exactly the one proposed by Sleeswyk for edge and mixed dislocations. Therefore, we conclude that this dislocation-based reaction can be the preferred mechanisms for the formation of twins, even though further atomistic simulation and experimental validation are required for a solid proof.

Fig. 4.6 Formation of twins in pure iron after impact loading, the comparison between experimental acquired EBSD analysis (a) and MD based orientation distribution (b) shows formation of twins in grain 1 and 2

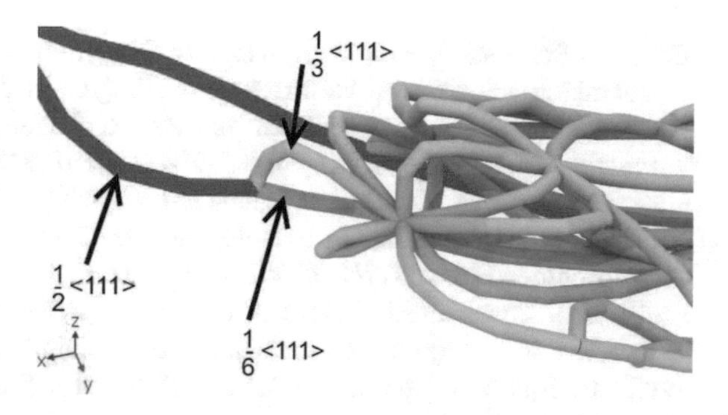

Fig. 4.7 Extraction of twin nucleation reaction. A 1/2 [111] dislocation dissociation on the tip of the newly formed twin results in a 1/3 [111] and 1/6 [111] partial dislocations which finally leads to twin growth

4.4 Conclusion

Deformation twinning was investigated in bcc iron after impact loading experiments. The temperature and grain size effects on twin formation were observed by optical microscopy and electron backscatter diffraction. According to the results, twinning is favored with increasing grain size and decreasing temperature. Furthermore, TEM observations obtained dislocation networks at the twin boundaries. An identification of the existing dislocation type was not successful, yet. As an alternative to limited experimental investigation, molecular dynamics simulation was introduced as suitable method. The MD simulations provide a good model to describe twin formation and twin growth. The results show a very good correlation to the experiments. The identified dislocation reactions stands also in good agreement to the theories known from literature, especially the pole mechanism. In further investigations, experimental and numerical studies are necessary to validate the numerically identified twin nucleation mechanisms with respect to the pole mechanism but also to other mechanisms described in literature.

Acknowledgements The authors thank the MDZWP e.V. for financial support of this study. The EBSD measurements were carried out by FEI SCIOS, funded by DFG large equipment funding (article 91b GG).

References

Bancroft D, Peterson EL, Minshall S (1956) Polymorphism of iron at high pressure. Journal of Applied Physics 27(3):291–298, DOI 10.1063/1.1722359

Christian JW, Mahajan S (1995) Deformation twinning. Progress in Materials Science 39(1):1 – 157, DOI 10.1016/0079-6425(94)00007-7

Cottrell AH, Bilby BA (1951) Lx. a mechanism for the growth of deformation twins in crystals. The London, Edinburgh, and Dublin Philosophical Magazine and Journal of Science 42(329):573–581, DOI 10.1080/14786445108561272

Ecke M, Schwarz F, Krüger L, Wilke M, Heyse H, Wendt U, Krüger M (2014) Charakterisierung von Verformungszwillingen in Meteoriten und dynamisch beanspruchten Eisenwerkstoffen mittels Rückstreuelektronenbeugung EBSD. Practical Metallography 51(11):765–784, DOI 10.3139/147.110317.

Gutierrez-Urrutia I, Zaefferer S, Raabe D (2010) The effect of grain size and grain orientation on deformation twinning in a fe–22wt.twip steel. Materials Science and Engineering: A 527(15):3552 – 3560, DOI 10.1016/j.msea.2010.02.041

Kadiri HE, Oppedal AL (2010) A crystal plasticity theory for latent hardening by glide twinning through dislocation transmutation and twin accommodation effects. Journal of the Mechanics and Physics of Solids 58(4):613 – 624, DOI 10.1016/j.jmps.2009.12.004

Krüger L, Wolf S, Martin S, Martin U, Jahn A, Weiß A, Scheller P (2011) Strain rate dependent flow stress and energy absorption behaviour of cast crmnni trip/twip steels. steel research international 82(9):1087–1093, DOI 10.1002/srin.201100067

Mahajan S (2002) Deformation twinning. In: Buschow KHJ, Cahn RW, Flemings MC, Ilschner B, Kramer EJ, Mahajan S, Veyssière P (eds) Encyclopedia of Materials: Science and Technology, Elsevier, Oxford, pp 1 – 14, DOI 10.1016/B0-08-043152-6/01809-X

Mendelev MI, Han S, Srolovitz DJ, Ackland GJ, Sun DY, Asta M (2003) Development of new interatomic potentials appropriate for crystalline and liquid iron. Philosophical Magazine 83(35):3977–3994, DOI 10.1080/14786430310001613264

Meyers MA, Vöhringer O, Lubarda VA (2001) The onset of twinning in metals: a constitutive description. Acta Materialia 49(19):4025 – 4039, DOI 10.1016/S1359-6454(01)00300-7

Neumann JG (1849) Über die kristallinische struktur des meteoreisens von braunau. Naturwissenschaftliche Abhandlungen Wien 3:45–56

Plimpton S (1995) Fast parallel algorithms for short-range molecular dynamics. Journal of Computational Physics 117(1):1 – 19, DOI 10.1006/jcph.1995.1039

Priestner R, Leslie WC (1965) Nucleation of deformation twins at slip plane intersections in B.C.C. metals. The Philosophical Magazine: A Journal of Theoretical Experimental and Applied Physics 11(113):895–916, DOI 10.1080/14786436508223953

Sleeswyk AW (1963) 1/2<111> screw dislocations and the nucleation of {112}<111> twins in the b.c.c. lattice. The Philosophical Magazine: A Journal of Theoretical Experimental and Applied Physics 8(93):1467–1486, DOI 10.1080/14786436308207311

Stukowski A (2009) Visualization and analysis of atomistic simulation data with OVITO–the open visualization tool. Modelling and Simulation in Materials Science and Engineering 18(1):015,012, DOI 10.1088/0965-0393/18/1/015012

Stukowski A, Albe K (2010) Extracting dislocations and non-dislocation crystal defects from atomistic simulation data. Modelling and Simulation in Materials Science and Engineering 18(8):085,001, DOI 10.1088/0965-0393/18/8/085001

Stukowski A, Bulatov VV, Arsenlis A (2012) Automated identification and indexing of dislocations in crystal interfaces. Modelling and Simulation in Materials Science and Engineering 20(8):085,007, DOI 10.1088/0965-0393/20/8/085007

Suzuki H, Barrett CS (1958) Deformation twinning in silver-gold alloys. Acta Metallurgica 6(3):156 – 165, DOI 10.1016/0001-6160(58)90002-6

Wang SJ, Sui ML, Chen YT, Lu QH, Ma E, Pei XY, Li QZ, Hu HB (2013) Microstructural fingerprints of phase transitions in shock-loaded iron. Scientific Reports 3:1–6, DOI 10.1038/srep01086

Zepeda-Ruiz LA, Stukowski A, Oppelstrup T, Vasily V (2017) Probing the limits of metal plasticity with molecular dynamics simulations. Nature 550:492–495, DOI 10.1038/nature23472

Chapter 5
Thermomechanical Cyclic Properties of P91 Steel

Władysław Egner, Stanisław Mroziński, Piotr Sulich, and Halina Egner

Abstract The paper presents the results of low-cycle fatigue tests on P91 steel specimens, carried out under constant strain and constant stress conditions, in several test temperatures. The analysis of the experimental data indicates the influence of fatigue testing conditions on the fatigue life predictions based on the Ramberg-Osgood relationship, while the increasing test temperature enhances the differences. In the next stage of the research the constitutive description of low cycle fatigue behavior of P91 steel is developed within the framework of thermodynamics of irreversible processes with internal state variables. The entropy production based damage model is applied to reflect the fatigue softening. The model is implemented into a numerical subroutine, and identified on the basis of the experimental data. A good agreement is obtained between simulations and experiment.

Key words: Low cycle fatigue · Damage · Constitutive modeling

5.1 Introduction

During the process of frequent starting-up and shutting-down of the power plant units, their components are subjected to cyclic changes in both mechanical load and temperature. P91 steel, which belongs to the 9-12% Cr steel grade members,

---

Władysław Egner · Piotr Sulich · Halina Egner
Institute of Applied Mechanics, Faculty of Mechanical Engineering, Cracow University of Technology, Al. Jana Pawła II 37, 31-864 Kraków, Poland,
e-mail: wladyslaw.egner@pk.edu.pl, piotrjansulich@gmail.com, halina.egner@pk.edu.pl

Stanisław Mroziński
UTP University of Science and Technology, Faculty of Mechanical Engineering, Bydgoszcz, Poland,
e-mail: stanislaw.mrozinski@utp.edu.pl

H. Altenbach et al. (eds.), *Plasticity, Damage and Fracture in Advanced Materials*, Advanced Structured Materials 121,
https://doi.org/10.1007/978-3-030-34851-9_5

exhibits excellent thermal properties, such as high creep resistance, good ductility, high resistance to thermal fatigue, good weldability, good thermal conductivity and low thermal expansion coefficient, good corrosion resistance and fracture toughness in water-steam and gas environments. For this reason this steel is very well suited for devices operating at temperatures up to 650°C, such as thick-walled pipes or forging for the construction of boilers, steam generators, nuclear reactors, and other responsible structural elements. The present investigation aims to determine the influence of testing conditions on the resulting fatigue material characteristics of P91 steel. Additionally, a constitutive description of the cyclic softening of P91 steel during low cycle fatigue at elevated temperatures is developed, based on the entropy production.

5.2 Experimental Tests

Experimental tests were performed on P91 steel specimens (Fig. 5.1) cut out of a boiler pipe (Sulich et al, 2017). The chemical composition of steel was 0.197 C, 0.442 Si, 0.489 Mn, 0.017 P, 0.005 S, 8,82 Cr, 0.971 Mo, 0.307 Ni, 0.012 Al, 0.017 Co, 0.036 Cu, 0.074 Nb, 0.004 Ti, 0.201 V and 0.02 W. Low-cycle fatigue tests were strain controlled, with constant total strain amplitude ε_{ac} and constant temperature in each test. The strain-controlled tests were carried out at five total strain levels ε_{ac} (0.6; 0.5; 0.35; 0.3; 0.25%). The tests at controlled stress (σ_a=const) were also carried out at five levels, determined on the basis of the experiment results at ε_{ac}=const. Since P91 steel does not exhibit the stabilization period, the corresponding stress amplitude levels were determined for the half of the fatigue life period. Three levels of temperature (20°C, 400°C and 600°C) were applied in each case of the mechanical control scheme (Mroziński et al, 2019). Experiments were performed on testing machine Instron 8502, equipped with heating chamber. Each load cycle was represented in 200 data points. The tested material generally exhibits an initial slight hardening during the very first cycles, the extend of it being dependent

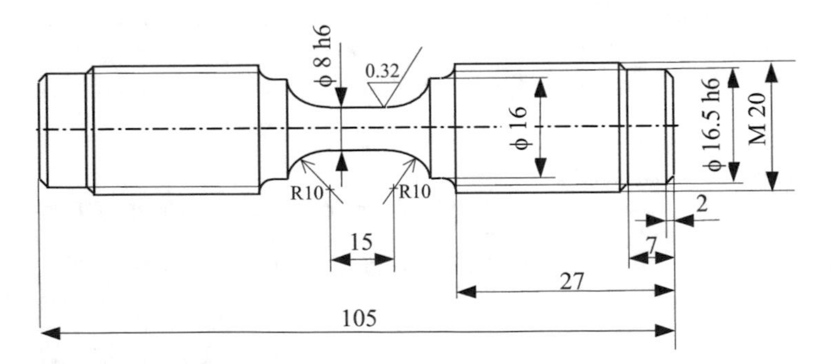

Fig. 5.1 Sample geometry

on the test temperature (decreasing with increasing temperature). After this very short consolidation phase the maximum stress on cycle continuously drops without a saturation period within the range of considered strain amplitudes, see Figs 5.2 and 5.3. The existing literature suggests various mechanisms responsible for the cyclic

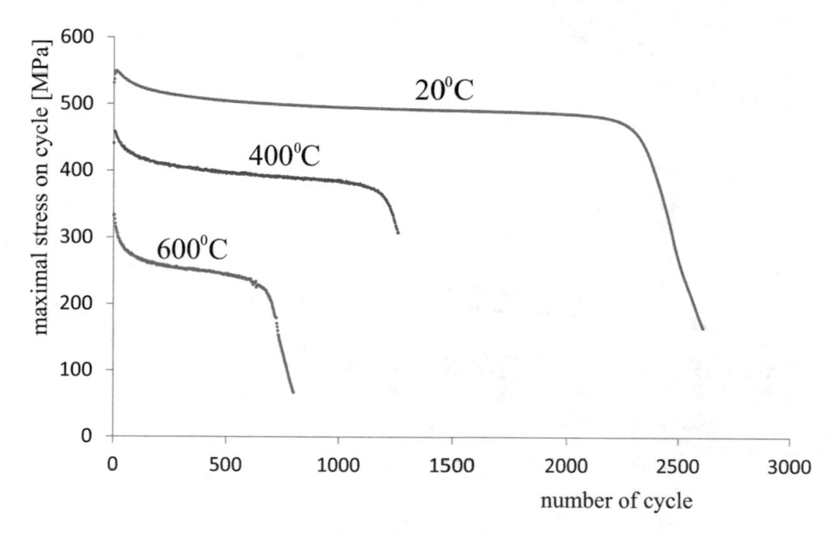

Fig. 5.2 Maximum stress on cycle versus number of cycle for strain amplitude ε_{ac}=0.5% and three test temperatures: 20°C, 400°C and 600°C

softening of the high strength steels group. In general this phenomenon is explained by the modification of dislocation structure and density, carbide morphology, density and chemical composition (Zhang et al, 2002), and fatigue damage.

5.3 Cyclic Properties

The analysis of the test results was performed with respect to the hysteresis loop parameters, required for the analytical Ramberg-Osgood relationship based description of the steel cyclic properties. The fatigue tests revealed variation of the considered parameters with the number of a load cycle, demonstrated by the modification of the hysteresis loop shape recorded during different stages of the fatigue life (Fig. 5.3). The analytical equation describing the relation between stress σ_a and plastic strain ε_{ap} has the following form:

$$\log\sigma_a = \log K' + n'\log\varepsilon_{ap} \tag{5.1}$$

The values of σ_a and ε_{ap} at corresponding strain ε_{ac} or stress σ_{ac} levels for each series of tests were determined using the least square method, by calculating the

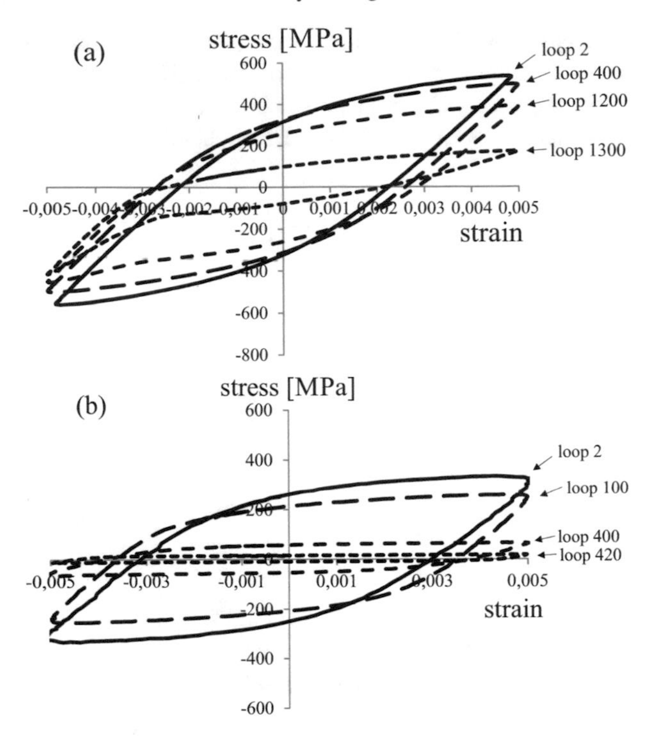

Fig. 5.3 Chosen hysteresis loops for test temperatures: (a) 25°C, (b) 600°C

coefficients and exponents of a regression line, according to Eq. (5.1). Figure 5.4 shows the stress-strain curves in the logarithmic scale, determined by approximation of the loop parameters (σ_a and ε_{ap}) at the load cycle corresponding to the half of the fatigue life. It is visible that the results slightly depend on the test conditions, and the effect is more pronounced at the elevated temperature - the differences in slope *n'* and absolute term *K'* are significantly higher at 600°C than at ambient temperature (20°C). However, when the energy $\Sigma\Delta W_{pl}$ dissipated at different loading levels is considered, the dependency on testing conditions is not observed. Figure 5.5 shows that the energy dissipated until failure very slightly depends on the loading mode, which is confirmed by similar absolute term and slope of the regression line:

$$\Sigma\Delta W_{pl} = \alpha_{\Delta W} \log N + K_{\Delta W} \tag{5.2}$$

Fatigue life calculations for structural components involve the operation of fatigue damage summation. The classical approaches based on the Palmgren-Miner hypothesis, e.g. the 'cycle after cycle' method, consist in a multiple repeated use (up to several thousand times) of the fatigue curve parameters. Therefore, even the slightest differences in these parameters may lead to significant discrepancies between the theoretical and experimental fatigue life results for steels that do not exhibit the stabilization period. In this context using the energy characteristics is beneficial in

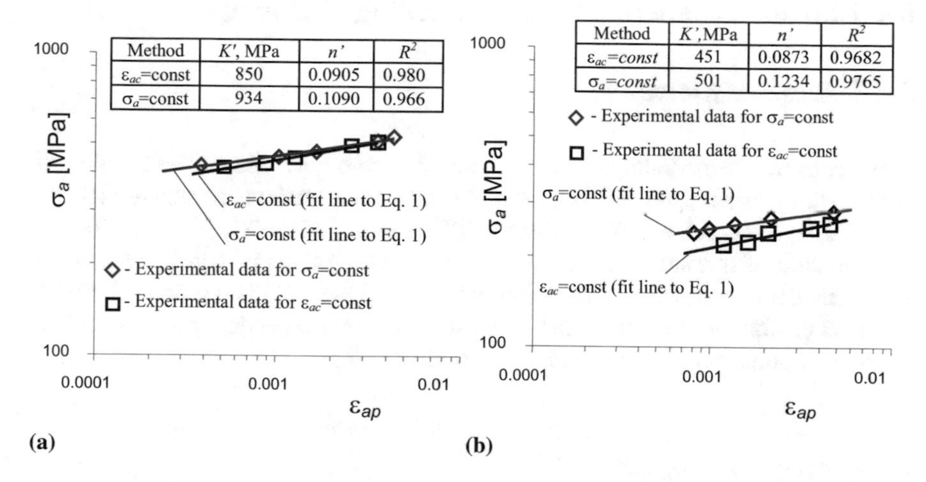

Method	K', MPa	n'	R^2
ε_{ac}=const	850	0.0905	0.980
σ_a=const	934	0.1090	0.966

Method	K',MPa	n'	R^2
ε_{ac}=const	451	0.0873	0.9682
σ_a=const	501	0.1234	0.9765

Fig. 5.4 Strain curves for P91 steel at different test temperatures: (a) $T = 20°C$, (b) $T = 600°C$

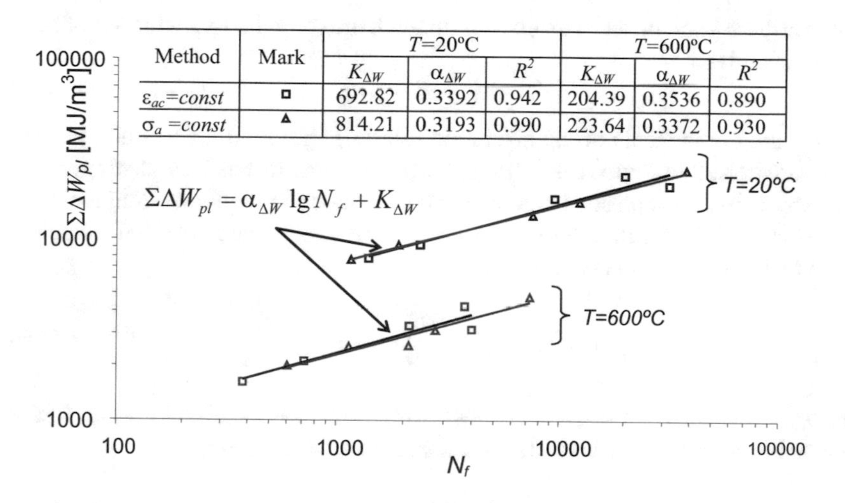

Method	Mark	T=20°C			T=600°C		
		$K_{\Delta W}$	$\alpha_{\Delta W}$	R^2	$K_{\Delta W}$	$\alpha_{\Delta W}$	R^2
ε_{ac}=const	□	692.82	0.3392	0.942	204.39	0.3536	0.890
σ_a =const	▲	814.21	0.3193	0.990	223.64	0.3372	0.930

Fig. 5.5 Energy $\Sigma\Delta W_{pl}$ dissipated until failure

the analysis of the low cycle fatigue process, in particular for the materials that do not exhibit stabilization.

5.4 Entropy Production Based Modelling of Damage

5.4.1 Basic Assumptions

In the constitutive modelling, the formalism of thermodynamics of irreversible processes with internal state variables, and the local state method are adopted (Sulich et al, 2018; Skrzypek and Kuna-Ciskał, 2003; Egner, 2012; Xie et al, 2019). The current state of a material is determined by certain values of some independent variables, called variables of state (observable or internal). The present model is based on the following assumptions: small strains, rate independent plasticity, mixed isotropic/kinematic plastic hardening, isotropic damage.

5.4.2 State Equations

For the elastic-plastic material exhibiting mixed hardening the following set of state variables is defined:

$$\{V_\alpha\} = \{\varepsilon^e_{ij}, \theta, \alpha_{ij}, r\} \tag{5.3}$$

where ε^e_{ij} are components of the reversible (elastic) strain tensor, θ is an absolute temperature in Kelvin degrees, α_{ij} corresponds to kinematic plastic hardening, while r is related to isotropic plastic hardening. The thermodynamic forces conjugated to state variables (5.3) result from the assumed form of the state potential, which is here the Helmholtz free energy ψ:

$$\sigma_{ij} = \rho \frac{\partial \psi}{\partial \varepsilon^e_{ij}}, \quad s = -\frac{\partial \psi}{\partial \theta}, \quad J_\alpha = \rho \frac{\partial \psi}{\partial V_\alpha} \tag{5.4}$$

where σ_{ij} is the stress tensor, s denotes entropy, and J_α are thermodynamic forces conjugated to internal state variables (back stress and drag stress):

$$\{J_\alpha\} = \{X_{ij}, R\} \tag{5.5}$$

The following hardening model is here adopted after Chaboche and Rousselier (1983a,b):

$$X_{ij} = \sum_{k=1}^{2} X^{(k)}_{ij}, X^{(k)}_{ij} = \frac{2}{3} C^{(k)} \alpha_{ij} \tag{5.6}$$

$$R = \sum_{i=1}^{2} R^{(i)}, R^{(i)} = Q^{(i)} \mathrm{e}^{-b^{(i)} r} \tag{5.7}$$

5.4.3 Cumulative Damage

Cumulative damage analysis plays a key role in life prediction of components subjected to cyclic loading. There are many concepts how to estimate degradation in the material microstructure, for example with the use of a scalar variable describing the loss of effective area from the initial to the damage state (Kachanov, 1958). When microvoids exhibit clearly directional nature, the tensorial variables of different rank are introduced (Murakami and Ohno, 1981; Krajcinovic, 1996; Murakami, 2012). Other concepts use change in density, variation in the cyclic plastic response, cumulative hysteresis energy etc. According to the second law of thermodynamics the development of damage must be accompanied by an increase in the internal entropy production, which may therefore be used as a damage indicator (Basaran and Nie, 2004). Total entropy S is a measure of disorder W in the system, through a relation established by Boltzmann (1898) (k_0 is the Boltzmann constant):

$$S = k_0 \ln W \tag{5.8}$$

The relation between the entropy per unit mass and the disorder parameter was derived by Basaran and Yan (1998):

$$W = \mathrm{e}^{w_0 s} \tag{5.9}$$

where w_0 is a parameter. In the present work damage is described after Basaran and Nie (2004) as a quantity proportional to the relative change in disorder:

$$D = D_{cr} \frac{\Delta W}{W_0} = D_{cr} \left(\mathrm{e}^{w_0 \Delta s} - 1 \right) \tag{5.10}$$

Symbol W_0 stands for the initial disorder of the material with specific entropy s_0, while D_{cr} stands for a material parameter. The change in specific entropy, Δs, can be calculated from the following expression (cf. Basaran and Nie, 2004):

$$\Delta s = \int_{t_0}^{t} \frac{\sigma_{ij} \varepsilon_{ij}^{p}}{\theta \rho} dt + \int_{t_0}^{t} \left(\frac{k^{\theta}}{\theta^2 \rho} |\mathrm{grad}\theta|^2 \right) dt + \int_{t_0}^{t} \frac{r^{\theta}}{\theta} dt \tag{5.11}$$

5.4.4 Evolution Equations

According to Eq. (5.10) the evolution of damage is governed by the following equation:

$$\dot{D} = D_{cr} w_0 \dot{s} \mathrm{e}^{w_0 \Delta s} \tag{5.12}$$

and

$$\dot{s} = \frac{\sigma_{ij} \dot{\varepsilon}_{ij}^{p}}{\theta \rho} \tag{5.13}$$

Evolution equations of internal state variables are obtained with the use of the normality rule applied to the dissipation potential:

$$F^p = f^p(\tilde{\sigma}_{ij}, \tilde{X}_{ij}, \tilde{R}) + \frac{3}{4}\left(\frac{\gamma^{(1)}}{C^{(1)}}\tilde{X}^{(1)}_{ij}\tilde{X}^{(1)}_{ij} + \frac{\gamma^{(2)}}{C^{(2)}}\tilde{X}^{(2)}_{ij}\tilde{X}^{(2)}_{ij}\right) \tag{5.14}$$

while f^p is von Mises-type yield function:

$$f^p(\tilde{\sigma}_{ij}, \tilde{X}_{ij}, \tilde{R}) = \sqrt{\frac{3}{2}(\tilde{s}_{ij} - \tilde{X}'_{ij})(\tilde{s}_{ij} - \tilde{X}'_{ij})} - \sigma_y - \tilde{R}. \tag{5.15}$$

and s_{ij}, X'_{ij} denote stress and back stress deviators, respectively. The dissipation potential is expressed in an undamaged fictitious configuration, with the use of the strain equivalence principle, in which

$$\tilde{\sigma}_{ij} = \frac{\sigma_{ij}}{1-D},$$

$$\tilde{X}_{ij} = \frac{X_{ij}}{1-D},$$

and

$$\tilde{R} = \frac{R}{1-D}$$

The following classical evolution equations are obtained on the basis of the normality rule:

$$\dot{\varepsilon}^p_{ij} = \frac{3}{2}\frac{\dot{\lambda}}{1-D}\frac{s_{ij} - X'_{ij}}{\sqrt{\frac{3}{2}(\tilde{s}_{ij} - \tilde{X}'_{ij})(\tilde{s}_{ij} - \tilde{X}'_{ij})}} \tag{5.16}$$

$$\dot{\alpha}_{ij} = \dot{\varepsilon}^p_{ij} - \alpha_{ij}\frac{\dot{r}}{1-D}(\gamma_1 + \gamma_2) \tag{5.17}$$

$$\dot{r} = \frac{\dot{\lambda}}{1-D} \tag{5.18}$$

5.5 Results

5.5.1 Numerical Implementation

The constitutive model is implemented into numerical subroutines by the use of the fully implicit backward Euler scheme and the Newton-Raphson method. In order to solve the problem numerically, the classical concept of elastic predictor/plastic corrector was applied (Sulich et al, 2017). Wolfram Mathematica software was used to generate the expressions for components of the Jacobian matrix, while the numerical subroutines were automatically generated in C ++ language by the use of AceGen software, see Fig. 5.6.

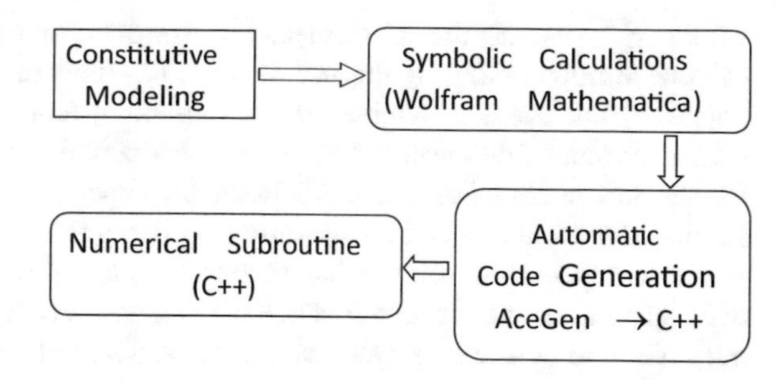

Fig. 5.6 Scheme of numerical implementation

5.5.2 *Identification Procedure*

The identification of model parameters for each isothermal test was performed with the application of SIMULIA-Isight package (Dassault, 2016). Two key components from the point of view of identification were used: "Data Matching", which offers the ability to calculate different error measures of two or more data sets (e.g. stresses obtained from the experiment, and stresses calculated numerically) and "Optimization" component, which allows for various methods of finding the minimum of a multi-variable function. The vector of material parameters contains components listed in Table 5.1. The vector of material parameters, bounded between their respective lower bounds and upper bounds, $P_i \in < L_i, U_i >$, was normalized:

$$\bar{P}_i = \frac{2P_i - (U_i + L_i)}{U_i - L_i}, \quad \bar{P}_i \in < -1, 1 > \tag{5.19}$$

At the beginning of the identification process the parameter bounding values were set arbitrarily, on the basis of the expected values for steel as an engineering material. The upper and lower bounds were then subjected to modifications if the optimisation procedure indicated different range. To find the set of optimal model parameters, the following error measure was used:

Table 5.1 Model parameters

E, ν, σ_y	Young's modulus, Poisson's ratio, yield stress
$C^{(1)}, C^{(2)}, \gamma^{(1)}, \gamma^{(2)}$	Parameters of kinematic hardening part
$Q^{(1)}, Q^{(2)}, b^{(1)}, b^{(2)}$	Parameters of isotropic hardening part
D_{cr}, w_0	Parameters related to entropy production (damage)

$$F_{obj}(\bar{P}_i) = w_1 \sum_{k=1}^{m} |\sigma_k(\bar{P}_i) - \sigma_k^{exp}| + m w_2 Max|\sigma_j(\bar{P}_i) - \sigma_j^{exp}|_{j=1,2,\ldots,m} \quad (5.20)$$

where σ_k^{exp} denote the experimental stress data, and $\sigma_k(\bar{P}_i)$ are the stress data calculated numerically by the use of current values of model parameters P_i. The objective function is a weighted sum (weights w_1 and w_2) of two components: the sum of absolute differences between the experimental and numerical data (stress), and the absolute maximal differences between the experimental data and the numerically simulated data. Such function allows to effectively reduce the maximum error ($w_2 \gg w_1$) or to adjust very well the data in the entire tested range ($w_1 \gg w_2$). In the case of fatigue modelling the identification of parameters is laborious, because it involves multiple time consuming cyclic simulations. The search time depends strongly on the selection of the starting point. In order to shorten the "distance" to the optimal solution, the following 5-step procedure was used to select the appropriate starting point as close as possible to the optimal solution:

1. First the elastic parameters (initial yield stress σ_y and elastic modulus E) were determined manually, taking into account the initial part of the first hysteresis loop.
2. Next, considering the entire first hysteresis loop, the approximate values of the kinematic hardening parameters were determined ($C^{(1)}, C^{(2)}, \gamma^{(1)}, \gamma^{(2)}$) assuming that the isotropic hardening in the first loop can be disregarded ($Q^{(1)} = 0, Q^{(2)} = 0$, $b^{(1)} = 0$ and $b^{(2)} = 0$).
3. Then, parameters related to isotropic hardening were searched on the basis of several selected hysteresis loops. As a result, approximate values of all material parameters were determined as the starting point for optimization.
4. In the next step damage related parameters were estimated, while other previously found coefficients were kept constant.
5. Finally, the identification of all material parameters was carried out again, but in a substantially reduced range around the starting point.

The exemplary results are given in Table 5.2.

5.5.3 Validation

The general procedure of constitutive modelling, numerical implementation and parameter identification was validated with the use of available experimental data.

Table 5.2 Model parameters for P91 steel at temperature 400°C

E [GPa]	σ_y [MPa]	$C^{(1)}$ [GPa]	$C^{(2)}$ [GPa]	$\gamma^{(1)}$ –	$\gamma^{(2)}$ –	$Q^{(1)}$ [MPa]	$Q^{(2)}$ [MPa]	$b^{(1)}$ –	$b^{(2)}$ –	D_{cr} –	w_0 $\left[\frac{\text{kgK}}{\text{J}}\right]$
184,7	281	128,4	92.8	688	499	-42,2	-40,9	0,21	2,84	6,20 E-8	583

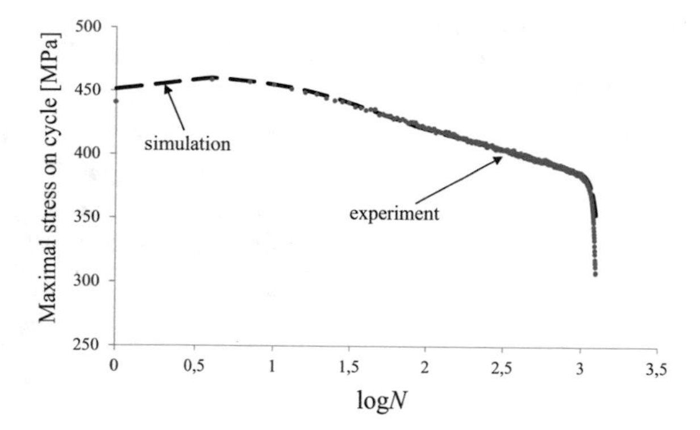

Fig. 5.7 Maximum stress on cycle versus logarithm of cycle number

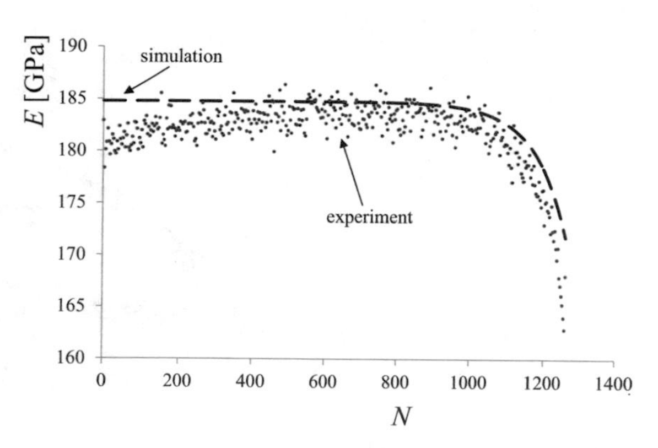

Fig. 5.8 Unloading modulus versus number of cycle

A good agreement was obtained between numerical simulations and experimental tests (see Figs. 5.7-5.9).

5.6 Conclusions

This study concerns the description of cyclic softening of P91 steel. The analysis presented in this work consists of five stages:

(1) experimental testing of the material at several test temperatures,
(2) constitutive modeling concerning variable temperature influence and the phenomenon of material cyclic softening,
(3) numerical implementation of the mathematical model,
(4) identification of model parameters, and
(5) validating the analysis by comparing the experimental and numerical results.

The analysis of experimental results for different testing conditions indicated the advantage of dissipation-based modelling over the approaches using stress- or strain-

Fig. 5.9 Maximum stress on cycle versus logarithm of cycle number

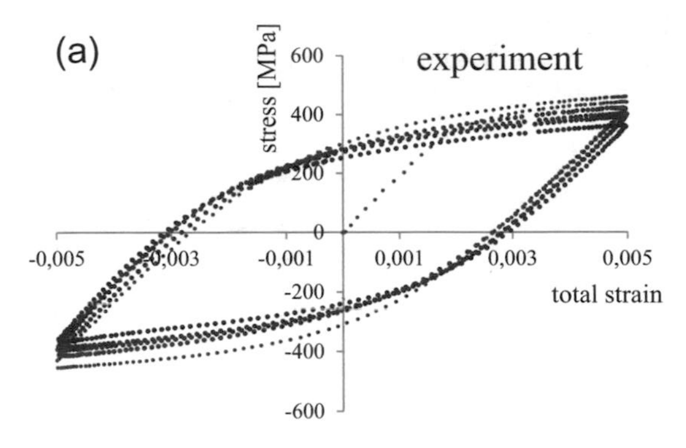

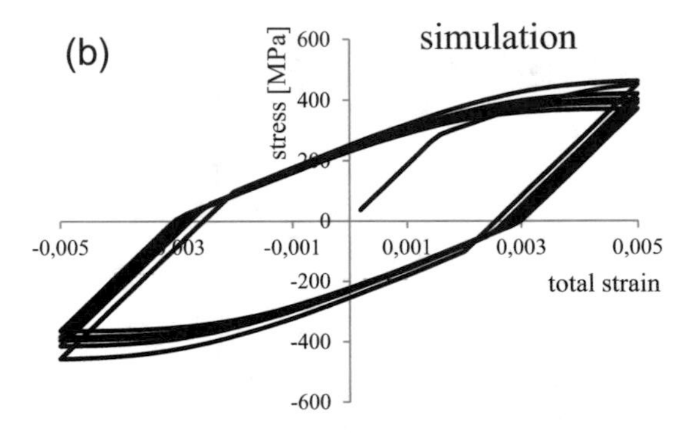

based measures, particularly for materials that do not exhibit cyclic stabilisation period. For this reason the description of damage softening was performed with the use of the entropy production to trace the evolution of material disorder. Such approach allows to efficiently describe the development of damage in both elastic and inelastic range of behaviour. To properly reflect the first two stages of material cyclic softening related to different dislocation populations occuring in the material microstructure, the drag stress was decomposed into two components. The first one (Eq. 5.7) is responsible for the initial nonlinear stage of rapid softening, while the second term allows to capture the quasi-linear softening. A 5-step identification procedure was applied to find the vector of optimal material parameters that provide the best fit into the experimental data for the applied objective function.

Acknowledgements This work was supported by the National Science Centre of Poland through the Grant No. 2017/25/B/ST8/02256.

References

Basaran C, Nie S (2004) An irreversible thermodynamics theory for damage mechanics of solids. International Journal of Damage Mechanics 13(3):205–223

Basaran C, Yan CY (1998) A Thermodynamic Framework for Damage Mechanics of Solder Joints. Journal of Electronic Packaging 120(4):379 – 384

Boltzmann L (1898) Lectures on Gas Theory. University of California Press, Berkeley, CA

Chaboche JL, Rousselier G (1983a) On the plastic and viscoplastic constitutive equations — Part I: Rules developed with internal variable concept. Int J Pressure Vessel Piping 105(2):153 – 158

Chaboche JL, Rousselier G (1983b) On the plastic and viscoplastic constitutive equations — Part II: Application of internal variable concepts to the 316 stainless steel. Int J Pressure Vessel Piping 105(2):159 – 164

Dassault (2016) Abaqus 6.14 - AP Isight 5.9. SIMULIA Abaqus Extended Products

Egner H (2012) On the full coupling between thermo-plasticity and thermo-damage in thermodynamic modeling of dissipative materials. International Journal of Solids and Structures 49(2):279 – 288

Kachanov LM (1958) O vremeni razrusheniya v usloviyah polzuchesti (time of the rupture process under creep conditions). Izvestiya AN SSSR Otd Mehn Nauk (8):26 – 31

Krajcinovic D (1996) Damage Mechanics. Elsevier, Amsterdam

Mroziński S, Egner H, Piotrowski M (2019) Effects of fatigue testing on low-cycle properties of P91 steel. International Journal of Fatigue 120:65 – 72

Murakami S (2012) Continuum Damage Mechanics. Springer, Berlin

Murakami S, Ohno N (1981) A continuum theory of creep and creep damage. In: Ponter ARS, Hayhurst D (eds) Creep in Structures, Springer, Berlin, pp 422–444

Skrzypek JJ, Kuna-Ciskał H (2003) Anisotropic elastic-brittle-damage and fracture models based on irreversible thermodynamics. In: Skrzypek JJ, Ganczarski AW (eds) Anisotropic Behaviour of Damaged Materials, Springer, Berlin, Heidelberg, Lecture Notes in Applied and Computational Mechanics, vol 9, pp 143–184

Sulich P, Egner W, Mroziński S, Egner H (2017) Modeling of cyclic thermo-elastic-plastic behaviour of P91 steel. Journal of Theoretical and Applied Mechanics 55(2):595–606

Sulich P, Egner W, Egner H (2018) Numerical analysis of thermomechanical low cycle fatigue. AIP Conference Proceedings 1922(1):150,005

Xie XF, Jiang W, Chen J, Zhang X, Tu ST (2019) Cyclic hardening/softening behavior of 316l stainless steel at elevated temperature including strain-rate and strain-range dependence: Experimental and damage-coupled constitutive modeling. International Journal of Plasticity 114:196 – 214

Zhang Z, Delagnes D, Bernhart G (2002) Anisothermal cyclic plasticity modelling of martensitic steels. International Journal of Fatigue 24(6):635 – 648

Chapter 6
Damage Identification Supported by Nondestructive Testing Techniques

Zbigniew L. Kowalewski, Aneta Ustrzycka, Tadeusz Szymczak, Katarzyna Makowska, and Dominik Kukla

Abstract Development of damage due to exploitation loadings was investigated using destructive and non-destructive methods in materials commonly applied in power engineering or automotive industry. The fatigue or creep tests for a range of different materials were interrupted for selected number of cycles or deformation level in order to assess a damage degree. As destructive methods the standard tensile tests were carried out after prestraining. Subsequently, an evolution of the selected tensile parameters was taken into account for damage identification. The ultrasonic, magnetic and novel optical techniques were used as the non-destructive methods for damage evaluation. The experimental programme also included microscopic observations. The results show that ultrasonic and magnetic parameters can be correlated with those coming from destructive tests. It is shown that good correlation of mechanical and selected non-destructive parameters identifying damage can be achieved for the materials tested. The results of damage monitoring during fatigue tests supported by contemporary optical techniques (Digital Image Correlation and Electronic Spackle Pattern Interferometry) proved their great suitability for effective identification of places of damage initiation.

The work additionally presents simulation of fatigue crack initiation for cyclic loading within the nominal elastic regime. It is assumed that damage growth occurs due to action of mean stress and its fluctuations induced by crystalline grain inhomogeneity and free boundary effect. The fluctuation fields in polycrystalline metal subjected to the mechanical loading inducing uniform mean stress and strain states development due to the material inhomogeneity related to grain anisotropy and inhomogeneity. The yielding process develops at the low mean stress level in some grains due to the local strain accumulation at their boundaries. These stress fluctuations,

Zbigniew L. Kowalewski · Aneta Ustrzycka · Dominik Kukla
Institute of Fundamental Technological Research, ul. Pawińskiego 5B, 02-106 Warsaw, Poland,
e-mail: zkowalew@ippt.pan.pl, austrzyc@ippt.pan.pl, dkukla@ippt.pan.pl

Tadeusz Szymczak · Katarzyna Makowska
Motor Transport Institute ul. Jagiellonska 80, 03-301 Warsaw, Poland,
e-mail: tadeusz.szymczak@its.waw.pl, katarzyna.makowska@its.waw.pl

H. Altenbach et al. (eds.), *Plasticity, Damage and Fracture in Advanced Materials*, Advanced Structured Materials 121,
https://doi.org/10.1007/978-3-030-34851-9_6

developing at a fraction of the macroscopic elastic limit, are the source of initial structural defects and microscopic plastic mechanisms controlling the evolution of defect assemble toward the state of advanced yielding. A mechanism responsible for damage accumulation during cyclic loading below the yield point remains elusive and requires classification. The analytical description is aimed at development of the consistent description of the microplastic state of material. The macrocrack initiation corresponds to a critical value of accumulated damage.

Key words: Creep · Damage · Fatigue · Optical methods (DIC, ESPI) · Nondestructive investigations · Magnetic and ultrasonic techniques · Microplasticity · Crack · Modelling

6.1 Introduction

In majority cases, degradation of a material has a local character and it is based on damage development leading to generation of cracks appearing around structural defects or geometrical notches. An identification of these areas and their subsequent monitoring requires a full-field displacement measurements performed on the objects surfaces. This chapter presents an attempt to use the Electronic Speckle Pattern Interferometry (ESPI) and Digital Image Correlation (DIC) for damage evaluation and its monitoring on specimens made of different materials subjected to static, monotonic or cyclic loading. Also effectiveness of other nondestructive techniques will be discussed. Among them one can indicate magnetic and ultrasonic methods.

Determination of material behaviour under various loading types can be reached by the use of different measurement techniques. The extensometer method is the most popular manner applied either in static or fatigue tests for strain components measurements versus time. This technique usually captures strain variations in a single direction defined at the beginning of experiment by the loading direction. The results collected in this way are important for typical engineering calculations carrying out in order to describe material behaviour in the uni-axial stress state conditions. For material examination under more complex loading, such data are insufficient, because many materials exhibit anisotropy of mechanical properties. Moreover, observations of damage zones and strain fields are not possible. Nowadays, the problems related to multiaxial stress/strain analysis may be effectively solved by the usage of modern, non-contact, optical methods, like DIC or ESPI for example.

DIC method, recommended for 3D measurements, is a stereoscopic technique that usually applies two CCD cameras, light sources and advanced software (GOM source). A mathematical description of DIC is available in the literature, Chu et al (1985) for example. The method requires an application of the special pattern represented by black dots on a grey background (Chu et al, 1985; Lord, 2009). Thanks to the markers defining the x, y and z coordinates, the method enables strain control up to the specimen fracture. DIC can use a pattern of rectangles or squares, that origins are directly selected for displacement/strain calculations. The results are presented in the form of full-field maps expressing strain distribution from beginning of a test up to the specimen fracture (Chu et al, 1985; Lord, 2009). In further analyses the DIC maps can be compared with FEA results (Toussaint et al, 2008; Gower and Shaw, 2010; Kamaya and Kawakubo, 2011) in order to validate material models or constitutive equations. Digital Image Correlation is recommended for static and fatigue tests conducted under various programs of loading. In the case of static experiments, the DIC system is more flexible than in tests under cyclic loading. This is due to the limited number of stages offered by DIC software to be recorded. Therefore, a concept to capture strain variations due to cyclic loading should be formulated before testing by means of the software targets or commands in C++ code.

The results coming from various research groups show that DIC can be used for examination of material behaviour under typical tensile and compressive tests (Forster et al, 2012), fracture toughness examinations (Durif et al, 2012), and experiments for determination of the geometrical imperfection effects, such as notches and holes (Lord, 2009). Data of monotonic tensile tests are usually represented by the full-field maps reflecting strain distributions up to the fracture appearance. The results of fracture toughness experiments present strain distributions close to the fatigue crack, and enables the stress intensity factor values to be determined.

Electronic Speckle Pattern Interferometry is the second well known optical technique for displacement and strain measurements on the specimens surface. This noncontact and noninvasive technique is used in the fields of experimental mechanics to obtain the displacement maps during stress testing. It is a type of the holographic interferometry based on the analysis of laser beam, distracted from the optically rough surface. The image of the specimen due to the reflected laser waves is detected by the CCD (charge-coupled device) sensor of a camera and then transferred to a computer.

ESPI method is based on the application of the elementary wave phenomenon - interference. The interference process involves two beams: the first one illuminates the specimen surface, and as the reflected beam interferes with a second one - the reference beam. Through the subtraction process of the speckle interferograms (before and after loading up to the defined levels), correlation fringes are obtained. Having them, a phase map can be generated. It represents a distribution of displacement components in each direction, separately (Andersson, 2013). Final full-field stress and strain phase maps are created as the result of mathematical operations under the fixed boundary conditions (measurement area dimensions) and the material parameters (elastic modulus and Poisson's ratio). The usage of this method during fatigue testing enables the location of the greatest concentration of stress-induced defects and allows with high accuracy to predict the damage initiation. The presented method is a promising tool for identification of the real discontinuities such as defects and fractures.

6.2 Material Degradation Assessments Supported by Digital Image Correlation

6.2.1 Digital Image Correlation – Short Characterization of the Aramis 4M System

In majority practical cases, the DIC system is equipped with two digital cameras, Fig. 6.1. Usage of the single camera only enables obtaining also two-dimensional results, and moreover, for this type of configuration a time needed to achieve the final result is much shorter than in the case of a two-camera system, however, we do not have opportunity to capture the results in the third direction. The dimensions of the measurement area that can be analyzed using the Aramis 4M system cover a range from 10 × 7 [mm] to 4000 × 2900 [mm].

Regardless of a type of DIC system, single or double-cameras, a performance of the tests planned must be preceded by the calibration procedure of a device using the calibration plate containing the characteristic reference points, Fig. 6.1c. The object must be unloaded during calibration stage. In order to ensure validity of the calibration, it is necessary to provide coordinate values to the points visible in the middle area of the calibration plate. They must be greater than the coordinates of the remaining points. The whole process is based on the principle of correlation and the method of looking for points of the same coordinate values. The procedure requires a definition of the contour for analysis and registration of its shape, Fig. 6.1a. For this purpose, the characteristic points of the analyzed layer are assigned by square

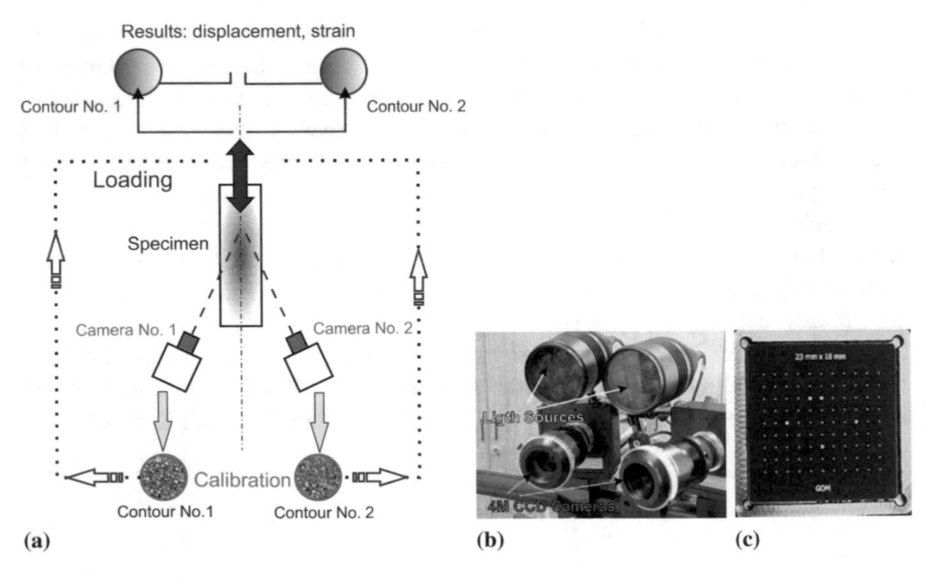

Fig. 6.1 ARAMIS system and its essential components: (a) general scheme of action (Toussaint et al, 2008), (b) main measurement module, (c) calibration plate

or rectangular areas (relatively small, e.g. 15 × 15 pixels) called "facets", Figs. 6.2 and 6.3.

Besides of the 15 × 15 pixel rectangular measurement areas, that can be changed in the DIC software, a segment of the gradient measuring zone is characterized by the presence of shared areas of the size of 2 × 2 pixels, Fig. 6.2. Into each of the rectangular areas a unique gray background with black points of any shape is assigned. It should be emphasized that in a research using DIC technique, it is necessary to select the shared zones. It should also be noted that sizes of "facets" affect the accuracy and speed of calculation. Increasing their dimensions decreases the accuracy of measurements, but on the other hand, it accelerates the time of the final result achievement. The rectangular areas are directly used in the analysis of displacement/strain components.

In the next stage the loading process is executed, during which the displacement components in the two- or three-dimensional area are determined. The values of displacement components are then used to calculate the strain/stress components in the form of full-field images. DIC system determines the coordinates in the two-dimensional (2D) system based on the reorientation of the rectangle/diamond centre of the facet, Fig. 6.3. The coordinates found using both cameras and the angle between their axes enable determination of the coordinates in the three-dimensional (3D) coordinate system. In the subsequent stages of analysis, a specific area of the layer with the gray-black background is identified. Its location can be used to find displacement. The initial stage (Stage 0) is assigned by the number "0", while the next stages by the subsequent numbers: "1", "2", "3", etc.

Difficulties occurring during DIC systems usage are mainly related to the preparation of the gray-black layer containing characteristic points located in the measuring zone of the specimen and the positioning of cameras, Fig. 6.1b, using a calibration plate, Fig. 6.1c.

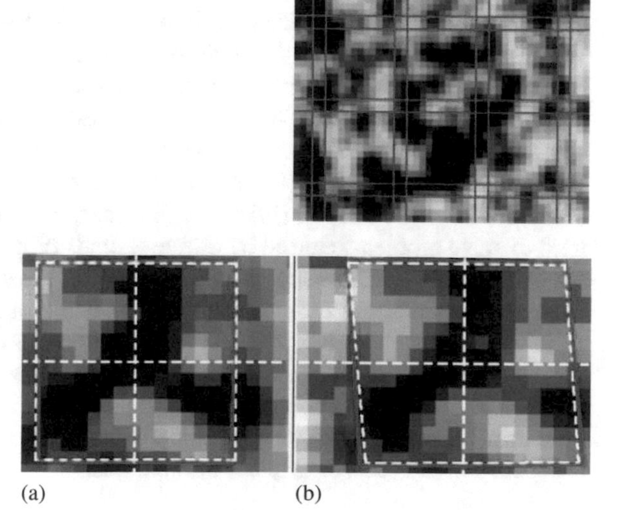

Fig. 6.2 A section of the measurement field with the facets arrangement defined by green lines (GOM, 2007)

Fig. 6.3 An example of an enlarged section of the analyzed area with a facet contour (green line) and a dashed line representing the relationship between the facets and strain: (a) before deformation, (b) after deformation (GOM, 2007)

DIC system designed for research in the field of large deformations can be successfully used to determine the Young's modulus, the Poisson's coefficient, and to identify the process of strengthening or weakening of the material in a wide range of plastic deformation. It is also applied in the study of materials with high ductility, exceeding 100%, and in determination of selected parameters of the fracture mechanics. In order to present high efficiency of DIC applications some representative results will be demonstrated in the next subsection.

6.2.2 Investigations of Material Behaviour Under Monotonic Tension Using Digital Image Correlation System

This section reports behaviour of engineering materials subjected to monotonic loading of specimen that had notches in the shape of "U" and "V". An influence of such geometrical imperfections on fatigue process is studied on the basis of the Wöhler diagrams, number of cycles to failure and fatigue notch factor. The results of experiments conducted by the use of Digital Image Correlation system called 4M Aramis are presented and discussed in detail. Tensile characteristic of the 41Cr4 steel, obtained by means of two independent techniques: standard extensometer and DIC, are compared. The experimental program provided the results that confirmed applicability of DIC technique for determination of typical mechanical properties. The equivalent full-field strain distributions are presented for specimens with and without the imperfections subjected to tension up to fracture. An influence of "U" and "V" notches on variations of tensile curve was evaluated.

6.2.2.1 Specimens with Notches

Notches are defined as geometrical imperfections of structural components. Their geometry is represented by the angle between their edges, radius and depth. Many results express values of stress concentration factor (SCF) obtained by analytical or numerical calculations. This enables the assessment of influence of stress concentration factor on stress values appeared in a notch tip. The second widely used method for determination of the SCF influence comes as a result of the development of Finite Element Analysis (FEA). In this case, advanced engineering software is used to solve problems caused by different kinds of geometrical discontinuities. In the case of this method many advantages can be stated, i.e. application of various 2D or 3D elements and analysis of macro- and micro-scales.

Its main drawbacks are material definition in the elastic-plastic state, as FEA requires definition of material hardening or softening reached not only in the uniaxial stress state, but also in a complex one. Despite of the use of modern, biaxial testing machines, the material examination under various combination of stress components is not easy and is still being under development.

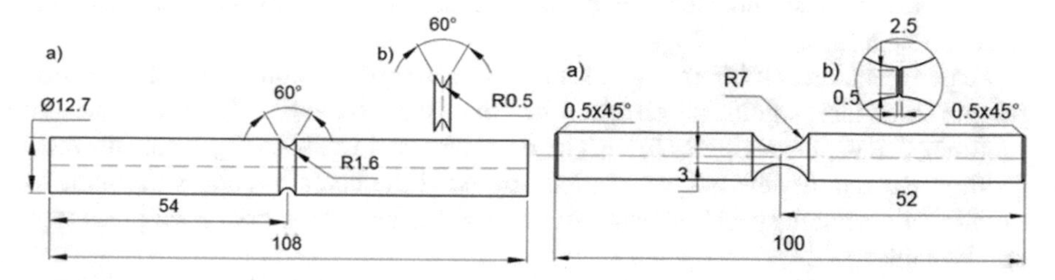

Fig. 6.4 Notched round specimens for fatigue testing (Fatemi et al, 2004; Pluvingae, 2001)

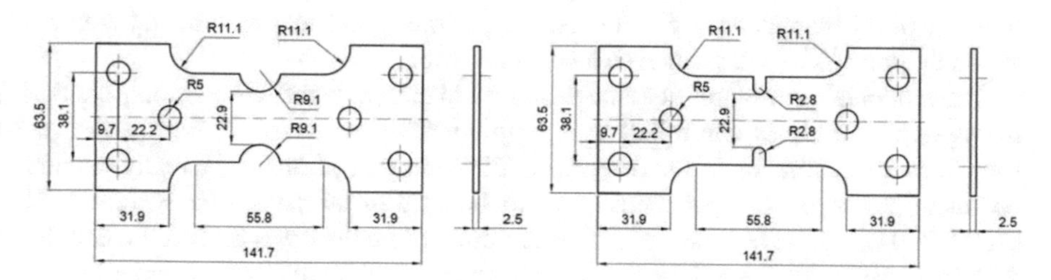

Fig. 6.5 Double-notched flat specimens used for examination of material behaviour under cyclic loading (Fatemi et al, 2004)

Notches can influence material behaviour and reduce the lifetime of components. They appear in drive shafts, engines, car bodies, turbines and others elements and constructions of complicated shapes. Among many features of notches, strongly related to their radius, angle, fillet at the tip, one can distinguish their size that can influence the concentration of stress/strain components, time to fracture and failure mode of the material.

Majority of specimens take a round or flat shape with macro- (Fig. 6.4) and micro-notches (Whaley, 1964; Maruno et al, 2003). The size of the geometric discontinuities is unique for each type of specimen, and depends on the details of the experimental procedure. In the case of round specimens, notches are machined around the major axis of the specimen (Figs. 6.4 and 6.5), but for flat specimens the geometric discontinuities are located on one (Whaley, 1964) or both sides of the specimen (Fig. 6.5).

6.2.2.2 Material Behaviour Affected by Notches and Holes – Previous Achievements

Essential efforts of many research groups (Wahl and Beuwkes Jr., 1930; Mazdumar and Lawrence Jr., 1981; Lanning et al, 1999; Milke et al, 2000; Fatemi et al, 2002, 2004; da Silva et al, 2012) are made to describe phenomenology of the influence of notches on material behaviour under various loading types up to fracture. As mentioned previously, notched specimens are usually designed by individual projects of research groups.

Experimental results from tests for determination of the influence of notches on fatigue show their significant effect on lifetime (Figs. 6.6-6.8). The differences in lifetimes due to various stress levels can be even equal to 80% (Fig. 6.8). Boronski (2007) assumed that the fatigue life determined at the notch bottom is the same as the life for a smooth specimen when strains waveforms for the both specimens type are the same.

The results presented in Fig. 6.7 illustrate variations of the fatigue notch factor for values of the stress ratio R within a range of 0.5÷1.0, due to the radius of the notch machined in cylindrical and flat specimens. In the case of flat specimens, an increase of this parameter is clearly visible. For the cylindrical specimen, the value of the fatigue notch factor achieved a constant level.

Experimental procedures are also designed to elaborate the Wöhler diagrams on the basis of the stress controlled tests, applying notches with different values of the stress concentration factor (Fig. 6.8). These investigations are usually carried out using round or flat specimens. As can be seen in the papers by Fatemi et al (2002, 2004), a great decrease of the cycle number to failure was achieved when the SCF increased three times. Distribution of experimental points from tests performed on flat specimens, with the notch corresponding to the stress concentration factor equal to 1.787, exhibited low agreement in comparison to the approximation by semi-logarithmic function. The reason of such disagreement is usually related to the specimens machining process and their mounting in the testing machines.

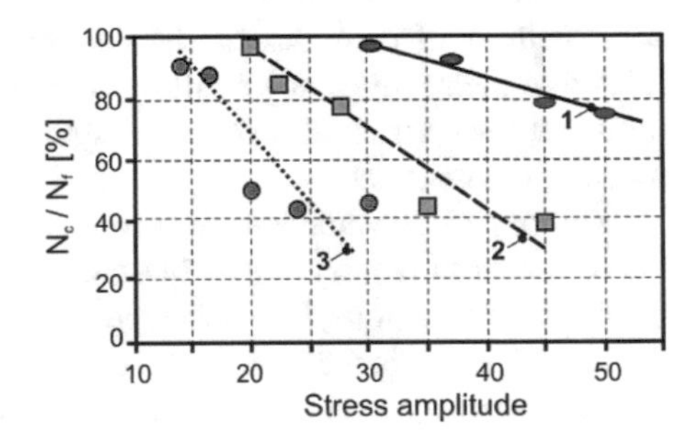

Fig. 6.6 Influence of notch radius on the proportion of cycle numbers to the initiation of the first crack (Nc) and fracture (Nf) at various stress amplitudes for the specimens: un-notched (1), notched with radius 1.5 mm (2); 6.35 mm (3); material: 24S-T4 aluminium alloy (Bennett and Weinberg, 1954)

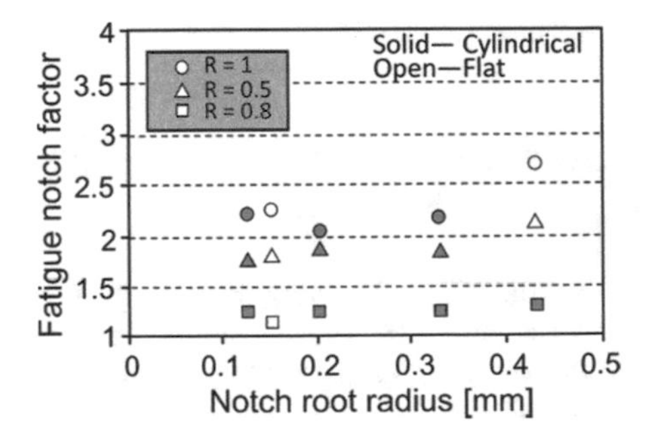

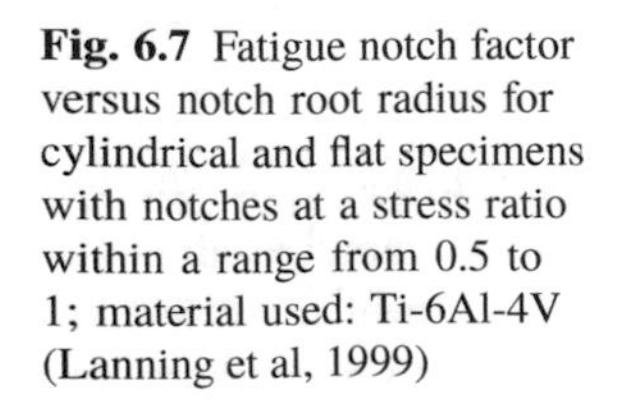

Fig. 6.7 Fatigue notch factor versus notch root radius for cylindrical and flat specimens with notches at a stress ratio within a range from 0.5 to 1; material used: Ti-6Al-4V (Lanning et al, 1999)

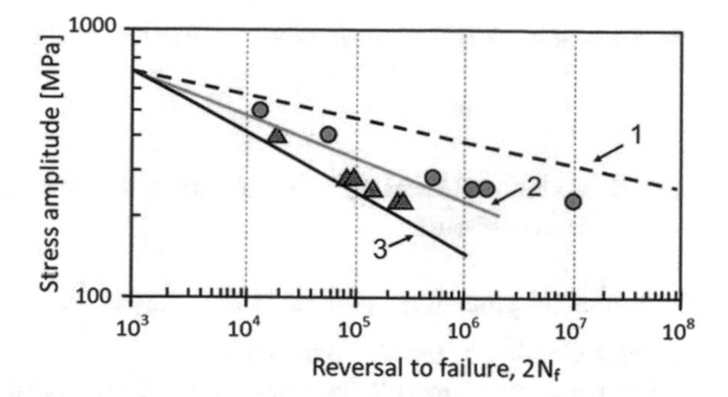

Fig. 6.8 Wöhler diagrams determined on the circumferentially notched round bars at stress concentration factors equal of 1.0 (1), 1.787 (2), 2.833 (3). Material: 1141 medium carbon steel micro-alloyed with Vanadium (Fatemi et al, 2002, 2004)

Other experiments were conducted to determine the zone of cracks initiation during fatigue (Fig. 6.9). In that case, the stress signal was used to control the testing machine. Both smooth and notched specimen geometries were applied (Qian et al, 2010). The application of smooth specimens in tests carried out at various stress levels allowed the initiation and propagation of cracks from the surface to the subsurface. In the case of notched specimens, the cracks appeared on the surface only.

Analytical and numerical calculations were concentrated on determination of stress gradients in the tip of the notch (e.g. Siebel and Stieler, 1955; Peterson, 1959; Milke et al, 2000). Their magnitudes were calculated using an equation representing dimensions of notches (Siebel and Stieler, 1955; Peterson, 1959; Pilkey, 1997), Neuber's theory (Neuber, 1961; Topper et al, 1967; Filippini, 2000; Neuber, 2001), and proportion of the stresses and nominal stress in the notch (Peterson, 1959; DuQuensay et al, 1986).

In FEM analysis, the influence of notches was calculated for specimens with narrow geometrical discontinuities like "V" or "U" (Pluvingae, 2001; da Silva et al, 2012; Andersson, 2013). The maximum stress in the tip of a notch (da Silva et al, 2012; Andersson, 2013) and the stress concentration factor were also determined

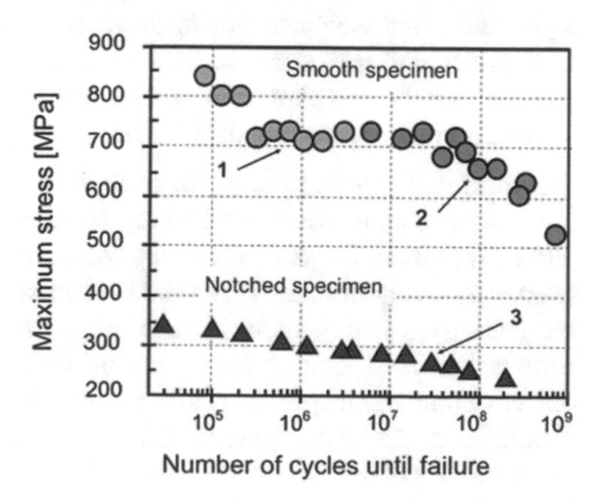

Fig. 6.9 Wöhler diagrams of the 40Cr steel, obtained for smooth and notched specimen with modes crack initiation: 1, 3 – surface; 2 – subsurface (Qian et al, 2010)

(da Silva et al, 2012) as well as an estimation of the fatigue strength (Andersson, 2013).

6.2.2.3 Material Behaviour Affected by Notches and Holes – Current own Achievements

Experimental procedure contained three stages:

1. determination of tensile characteristic;
2. investigation of the full-field strain distribution close to "U" and "V" notches;
3. examination of the influence of dimensions of "U" or "V" notches and their interactions on material behaviour.

All tests were carried out at room temperature using servohydraulic 8802 Instron and electro-dynamic Electropuls E10000 Instron testing machines. The Aramis 4M Digital Image Correlation system was applied to determine distribution of the strain components. Before testing, DIC device was calibrated. In the testing stages focused on determination of mechanical properties and material behaviour with the assistance of geometrical imperfections and various tensile rates, an extensometer and the 4M Aramis were used simultaneously. The assessment of DIC system used for determination of mechanical properties was made on the basis of a comparison of data obtained by the extensometer and by virtual tensometers defined in DIC software.

Application of Digital Image Correlation technique required the following stages:

1. adjustment of the distance between two cameras, its angle indicated in guidelines of technical data;
2. positioning the 4M Aramis with respect to the centre of the measurement zone;
3. selection of a calibration plate. that should be chosen on the basis of dimensions of the region for which the displacement is considered to be determined;
4. performing calibration procedure by application of the plate which takes various orientations in the 3D coordinate system, and recording its positions;
5. mounting the specimen, having an artificial measuring zone represented by black dots stochastically arranged on the grey layer, in grips of the testing machine;
6. capturing the first photo and establishing it as the reference one for displacement determination and strain calculations.

Technical data (GOM folders) for calibration of DIC system were delivered by the producer and contained the following features: measuring volume (height, length, width), minimum length camera support, distance ring (it enable to change a camera lens), measuring distance (from the central section of DIC device to the centre of the measuring zone), slider distance (determined by two technical points on the cameras), camera angle (it determines the centre of the measuring zone), calibration object (a plate with special regular markers having determined coordinates, which should be identified by cameras during the calibration process).

In the case of the experiments, the 4M Aramis was used at the following technical parameters:

- initial measuring zone determined on the specimen tested 25 × 10 × 3 [mm];
- calibration plate 25 × 18 [mm];
- slider distance 37.5 [mm];
- camera lens 75 mm + slider distance;
- camera angle 25°;
- sampling rate 2 photos/s.

Comparison of tensile curves obtained by the use of the flat specimen and the extensometer as well as DIC system is presented in Fig. 6.10. The axial stress was calculated as a proportion of the axial force to the cross-section of the measurement zone. Values of the axial strain were calculated for the same base.

Digital Image Correlation device and extensometer technique were simultaneously used in the monotonic test. For DIC calculations, two virtual tensometers were selected to define the gauge length for determination of the axial strain. A comparison of the results captured from the extensometer and DIC technique expressed a good agreement in the stress-strain curves from the beginning of testing up to the ultimate tensile strength occurrence. Differences in the last section of those characteristics are related to the appearance of neck, which was located close to the extensometer edge.

Mechanical properties calculated on the basis of data obtained by means of both techniques, are listed in Fig. 6.10. Differences between tensile characteristics are very small, indicating DIC system to be recommend not only for capturing strain maps (Fig. 6.11), but also for determination of typical mechanical parameters.

The main advantage of DIC system is presented in Fig. 6.11, which shows variations of the axial strain distribution at various stages of tension. On the basis of these results, the damage zone can be observed up to the specimen fracture (Fig. 6.11b-6.11d). It should be also mentioned that the distribution of the major strain (black vectors) at the beginning of tensile testing enables the assessment of specimen mounting quality in the testing machine (Fig. 6.11a).

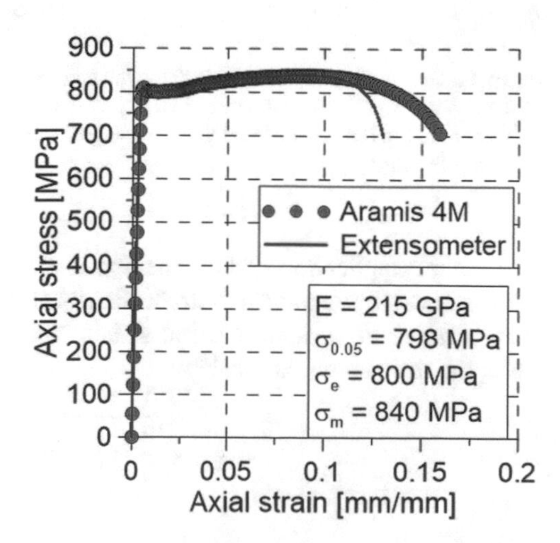

Fig. 6.10 Comparison of tensile characteristics determined by the use of the extensometer and the Aramis 4M. Material: high strength steel

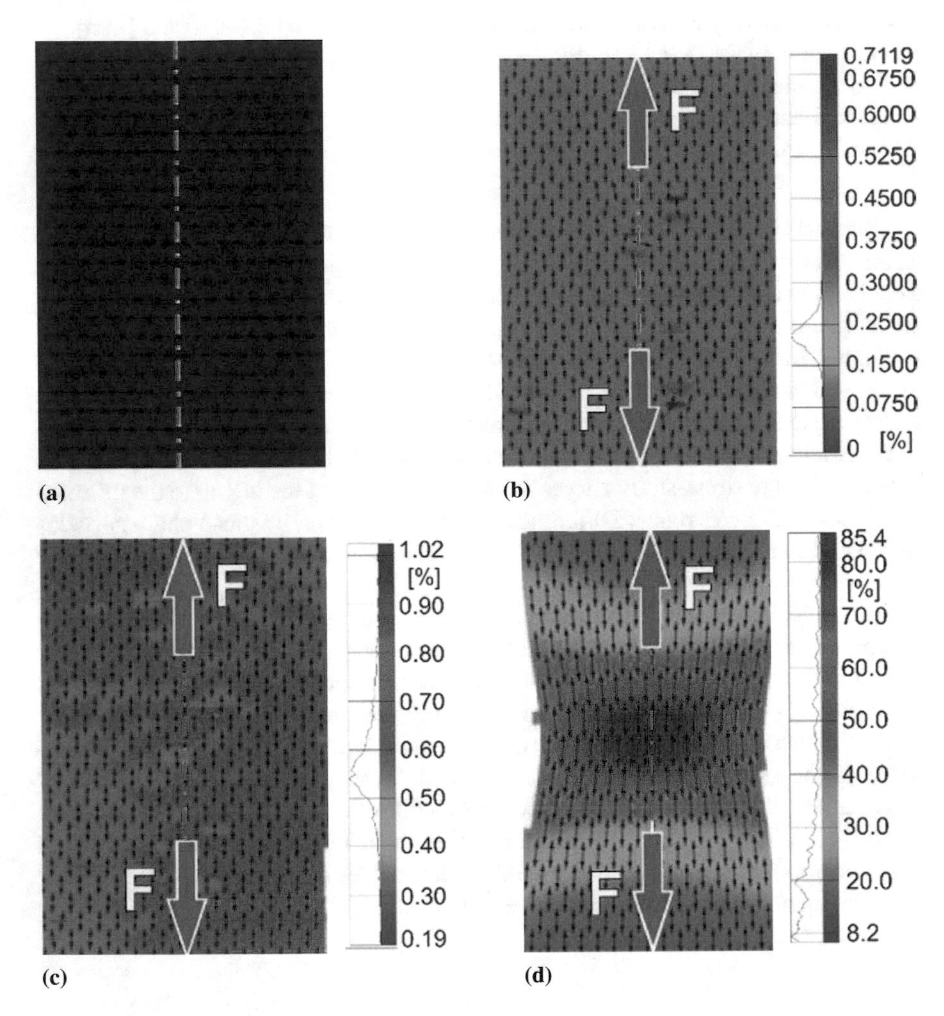

Fig. 6.11 Distribution of axial strain in the flat specimen under monotonic tension at various stages of the testing: (a) specimen mounted in the testing machine before the experiment; (b) stage corresponding to the proportional limit; (c) stage corresponding to the yield point; (d) stage before specimen fracture

Another application of DIC method is illustrated in Fig. 6.12. The subsequent images show the equivalent strain distributions for the specimen with the "U" - shaped notch at various material states. The 4M Aramis system calculates them using the following relationship:

$$\varepsilon_{\mathrm{eq}} = \sqrt{\frac{2}{3}\left(\varepsilon_{1_{\mathrm{true}}}^2 + \varepsilon_{2_{\mathrm{true}}}^2 + \varepsilon_{3_{\mathrm{true}}}^2\right)} \tag{6.1}$$

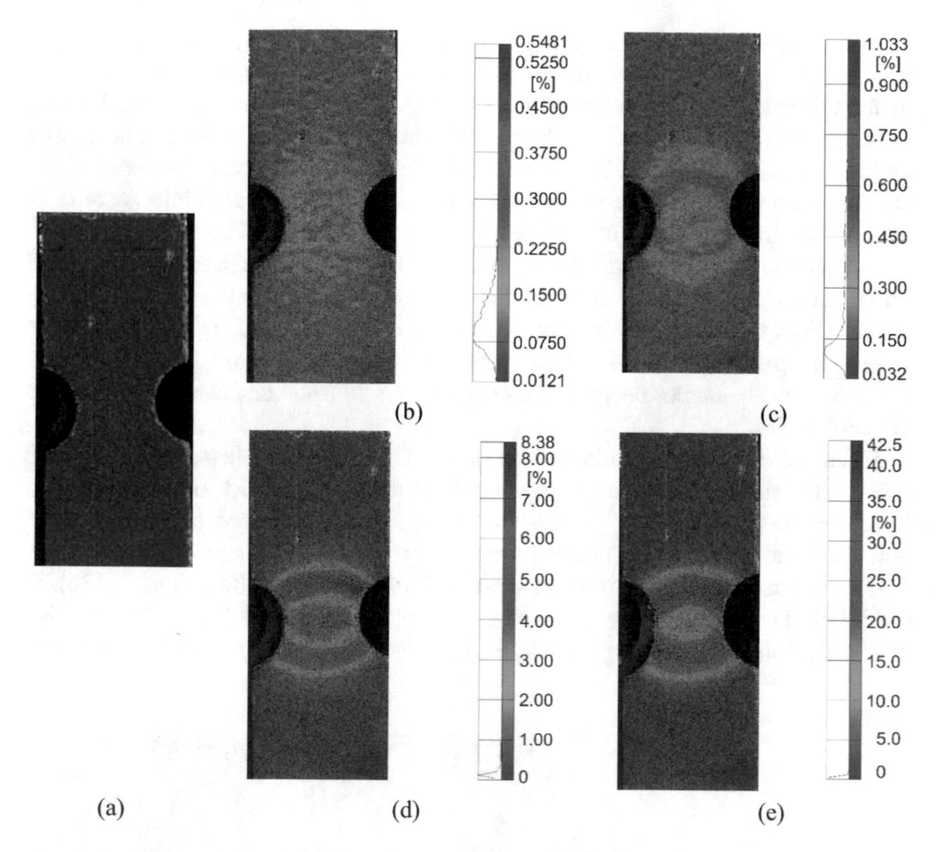

Fig. 6.12 The equivalent strain distribution in the "U"-notched specimen: (a) before loading, (b) elastic state, (c) initial phase of plastic state, (d) localization of damage, (e) damage development close to fracture. Material: aluminium

where: ε_i represents the major true strain components, $i = 1, 2$ and 3. The major strain components are defined by the following equation

$$\varepsilon 1_{\text{true}} = \ln(1 + \varepsilon)$$

for which $\varepsilon = \Delta l / l_0$ is the engineering strain, where l_0 - gauge length, Δl - elongation. Equation (6.1) represents the stress state in the strain coordinate system. It can be found on the basis of theory of plasticity (Olszak, 1965; Chen, 2004; Westergaard, 2014). It should be emphasized that Eq. (6.1) can be used when the straining is proportional, i.e. the constant ratios of $d\varepsilon 1_{\text{true}} / d\varepsilon 2_{\text{true}} / d\varepsilon 3_{\text{true}}$ are fulfilled.

Both initial images show the specimen at the beginning of the test, i.e. the referential stage with zero loading and in the elastic state (Fig. 6.12a, b). The elastic-plastic state, shown in Fig. 6.12c, expresses the equivalent strain distribution and indicates the deformation zone located close to the centres of the "U"-notches being initiators of the maximum strain.

These arc-shaped deformation regions appear where the growing damage occurs. It is strongly evidenced in the following stages of specimen tension, presenting the material behaviour before fracture (Fig. 6.12d, e). As it can be noticed in Fig. 6.12c-e, the strain distribution is expressed by the zone in form of an arc at the test beginning up to the specimen fracture. It shows that besides of the strain concentration, which can be captured by DIC, the shape of strain fields should be taken into account as data for the further analysis of the effects related to "U"-shaped notch.

DIC system was also applied for determination of the influence of geometrical imperfection dimensions in form of the "U"-shaped (Figs. 6.13, 6.14) and "V"-shaped (Figs. 6.15, 6.16) on the material fracture. The depth of the notches was the same and equal to 1.3 mm. The radii of the "U"-notches were of 0.75 mm, 1.5 mm and 2.5 mm. The angles between the edges of the "V"-notches were expressed by: 10°, 60° and 90°.

Effects resulting from the dimensions of the "U" and "V" notches were determined based on variations of the equivalent strain isolines, Figs. 6.13, 6.15 (Szymczak, 2018). In the case of the "U" multi-notched specimen, the interaction between the notches became more significant with an increase of theirs radius (Fig. 6.13b). The results achieved for further tension also reflected this effect (Fig. 6.13c). It disappeared when the damage zone became greater (Fig. 6.13d). The cracks can also be followed on the basis of typical photos from the CCD cameras of DIC system (Fig.

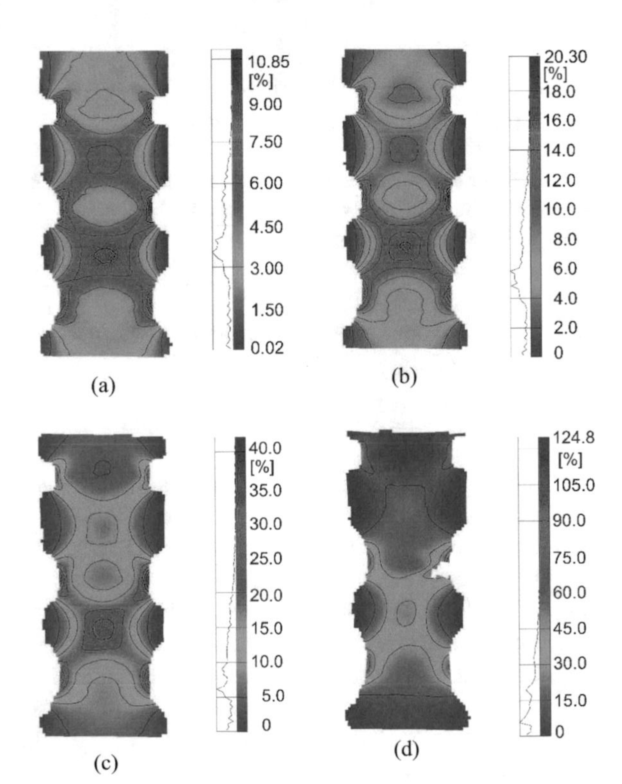

Fig. 6.13 The equivalent strain distribution in the U-shaped, multi-notched specimen at various stages of monotonic tension: (a), (b), (c) elastic-plastic state; (d) elastic-plastic state before fracture. Material: the 41Cr4 steel

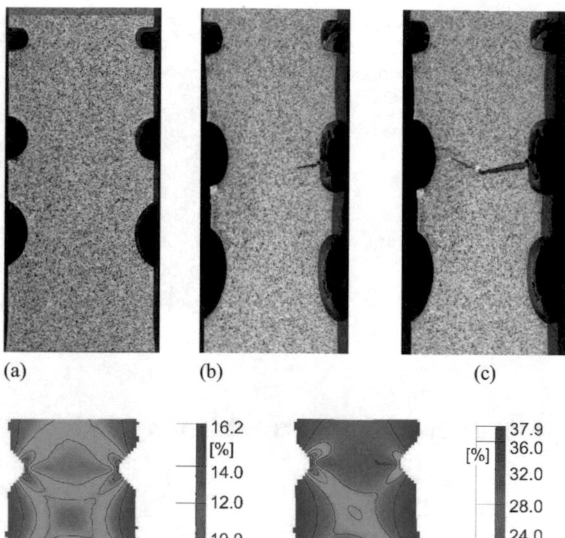

Fig. 6.14 The three stages of the "U"-notched specimen during monotonic tension: (a) before loading, (b) crack appearing, (c) crack growing

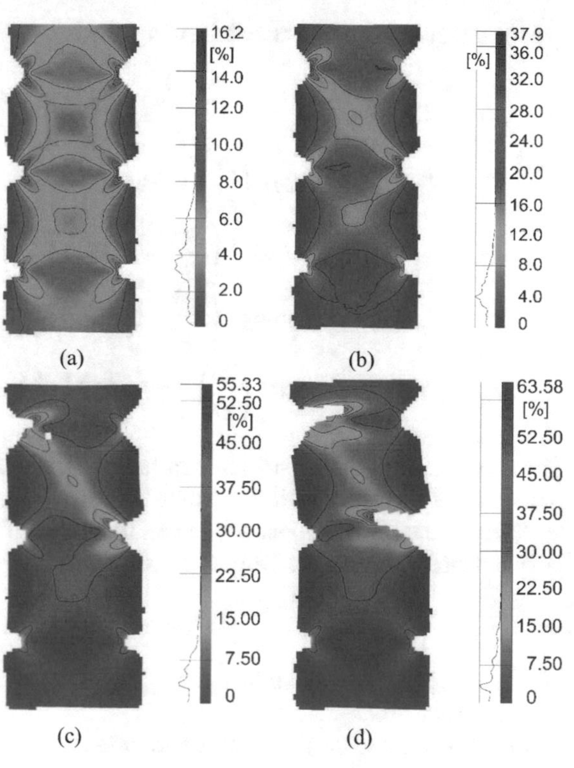

Fig. 6.15 The equivalent strain distribution in the multi-notched V specimen at various stages of monotonic tension: (a), (b) elastic-plastic state; (c), (d) elastic-plastic state before fracture. Material: the 41Cr4 steel

6.14). This is very important for the final stage of testing, because crack growing causes fracturing measurement pattern and, therefore, the digital correlation is not possible to be done (Figs. 6.13d, 6.14b, c).

Assessment of the effects of notches has been carried out using analytical equations for the stress concentration factor and the maximum stress (Pilkey, 1997). Their form for the multi-notched "U" specimens is noted in equation set (6.2), where: K_{tU} is the stress concentration factor, r is the notch radius, D - is specimen width, and L

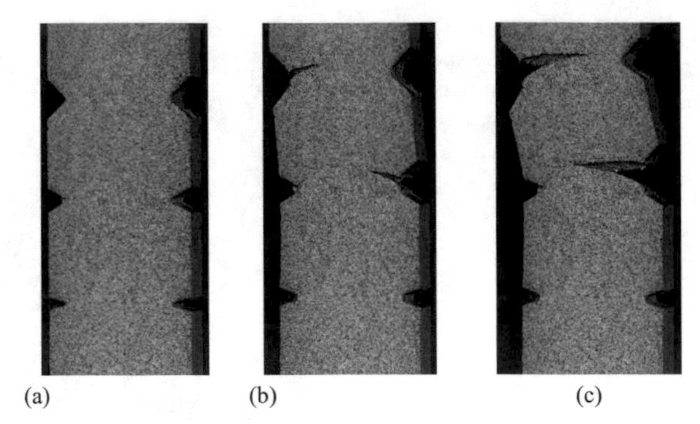

Fig. 6.16 The three stages of the "V" multi-notched specimen during monotonic tension: (a) before loading, (b) at crack initiation, (c) at crack growth

- is the distance between the notches, Fig. 6.17.

$$
\begin{aligned}
K_{tU} &= C_1 + C_2\left(\frac{2r}{L}\right) + C_3\left(\frac{2r}{L}\right)^2 + C_4\left(\frac{2r}{L}\right)^3,\\
C_1 &= 3.1055 - 3.4278\left(\frac{2r}{D}\right) + 0.8522\left(\frac{2r}{D}\right)^2,\\
C_2 &= -1.4370 - 10.5053\left(\frac{2r}{D}\right) - 8.7547\left(\frac{2r}{D}\right)^2 - 19.6237\left(\frac{2r}{D}\right)^3,\\
C_3 &= -1.6753 - 14.0851\left(\frac{2r}{D}\right) + 43.6575\left(\frac{2r}{D}\right)^2,\\
C_4 &= 1.7207 + 5.7974\left(\frac{2r}{D}\right) - 27.7463\left(\frac{2r}{D}\right)12 + 6.0444\left(\frac{2r}{D}\right)^3
\end{aligned}
\tag{6.2}
$$

Variations of the stress concentration factor versus notch radius expressed a linear reduction of its value with the radius increase, Fig. 6.18a. The same course was noticed for the maximum stress, as can be seen in Fig. 6.18a. This magnitude has been reached applying the following formula:

$$\sigma_{max} = K_{tU}\sigma_{nom} \tag{6.3}$$

where: $\sigma_{nom} = F/A_0$ – nominal stress, A_0 – area of specimen cross section for the measurement zone.

Looking at these results, it is easy to indicate the notches which are the dominant geometrical imperfections for the crack appearance, i.e. having a radius of 0.75 mm. This fact has not been confirmed by the DIC results in the final stage of tension (Fig.

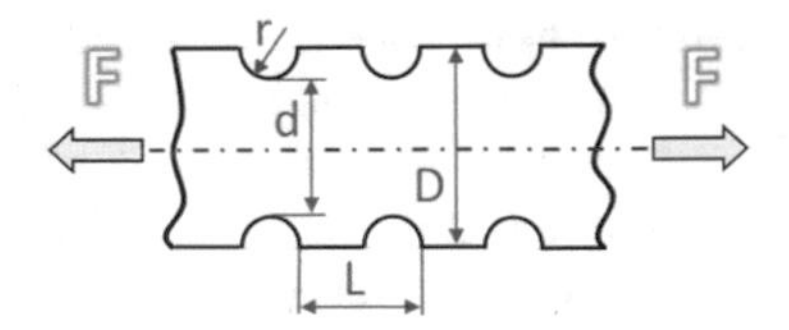

Fig. 6.17 "U" multi-notched specimen with the dimensions (Pilkey, 1997)

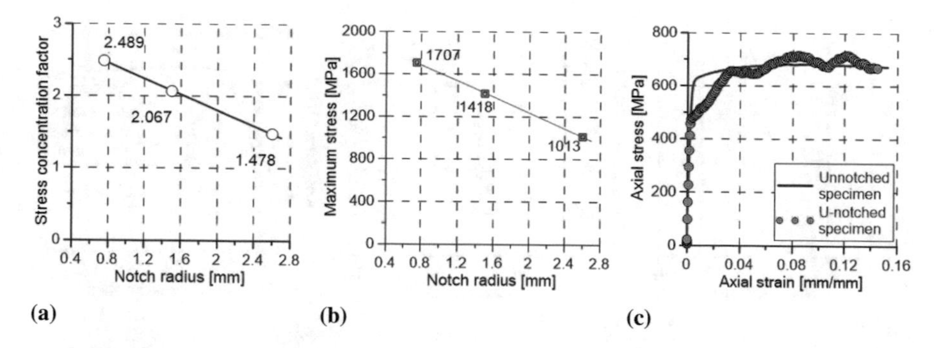

Fig. 6.18 The results for the multi-notched "U" specimen: (a) and (b) stress concentration factor and maximum stress as a function of notch radius, respectively; (c) tensile curve variations calculated by employing the cross section of the specimen

6.13), where the main crack occurred in the middle notch, for which the calculated maximum stress was lower (close to 300 MPa) than for the smallest one radius considered.

The effects resulting from the "V" notches presence can also be captured by means of DIC, Fig. 6.15 (Szymczak, 2018). For this type of geometrical imperfections, the full-field equivalent strain distributions close to the tip of the notches and the entire measurement section can be observed (Fig. 6.15b). Besides the different values of "V" notch angle, an indication of the main concentrator for the maximum strain is difficult. The 30° and 60° notches appear to have a very similar influence on the strain distribution at the beginning of the test. Further tension leads to an increase of the strain level in both notches located diagonally, which appear to be the main reason for initiation of damage zone (Fig. 6.15c). As a consequence, the equivalent strain increased, generating a fracture in the middle notch and then in the diagonal one (Fig. 6.15d).

For this type of geometrical imperfections, the stress concentration factor and the maximum stress have been calculated by the use of the following equation set (Pilkey, 1997):

$$\begin{aligned} K_{tV} &= C_1 + C_2\sqrt{K_{tU}} + C_3 K_{tU}, \\ C_1 &= -10.01 + 0.1534\alpha - 0.000647\alpha^2, \\ C_2 &= 13.60 - 0.2140\alpha + 0.000973\alpha^2, \\ C_3 &= -3.781 + 0.07873\alpha - 0.000392\alpha^2, \end{aligned} \tag{6.4}$$

where, K_{tV}, K_{tU} are the stress concentration factors for the "V" and "U" notches, respectively; C_i are the coefficients, and α is the angle between the edges of the notches. The maximum stress has been found applying the following relationship

$$\sigma_{max} = K_{tV}\sigma_{nom}$$

The results express a small lowering of the stress concentration factor and the maximum stress with decreasing of the notch angle, Fig. 6.19. These data well

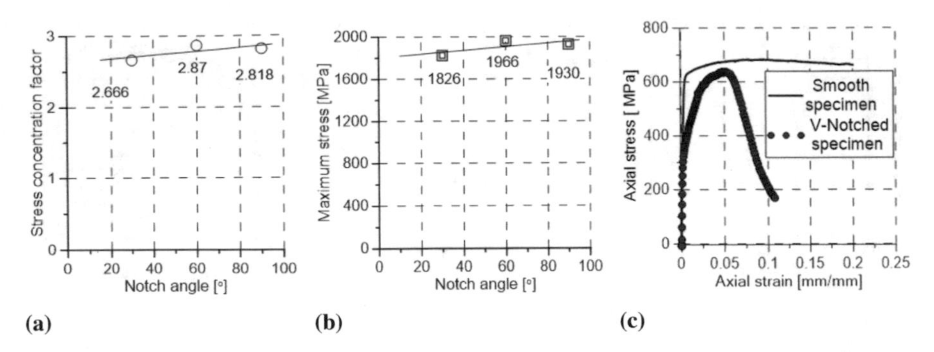

Fig. 6.19 The results for the "V" multi-notched specimen: (a) and (b) stress concentration factor and maximum stress as a function of notch angle, respectively, (c) tensile curve variations with respect to dimension of the cross section of the specimen

reflect the final stage of the multi-notched "V" specimen tension (Fig. 6.15d). Small differences between the values of the stress concentration factor and the maximum stress at various notch angles show that all of the "V" notches examined can also be treated as the potential places where the damage zones can occur.

The influence of the "V" notch on material straining during tension is illustrated in Fig. 6.19c presenting a comparison of tensile curves determined by the use of smooth and "V" multi-notched specimens. It is easy to notice that such geometrical imperfections caused the 50% reduction of elongation and 30% lowering of the yield point.

6.2.2.4 Damage Development Analysis Under Fatigue Conditions

Fatigue investigations were carried out on the MTS 810 testing machine, on plane specimens. These tests included specimens with three types of nickel alloy structure and with the layer thickness of 20 and 40 μm. Due to the continuous record of displacement maps, these tests were performed at high stress amplitude values equal to 600 MPa and 650 MPa, in order to reduce the test time to several hours. The loading frequency was 20 Hz, and the image was recorded every 5 seconds, i.e. the map was recorded every 100 cycles. Table 6.1 presents images from selected fatigue cycles obtained on a coarse grained specimen with a layer thickness of 20 μm. The amplitude of the alternating stress was 600 MPa. The specimen broke after 46 364 cycles, which is why the images recorded at the end of the specimen were compacted. For this specimen, the first signs of localization are visible after 45 thousands loading cycles. The increase in deformation at this stage is connected with the formation of a crack in the aluminide layer, which propagates into the material until the cohesion is entirely lost. On this basis, the moment of crack formation can be estimated, although the detection threshold is lower than that in the technically comparable ESPI method

Table 6.1 Images showing strain distribution maps captured by DIC

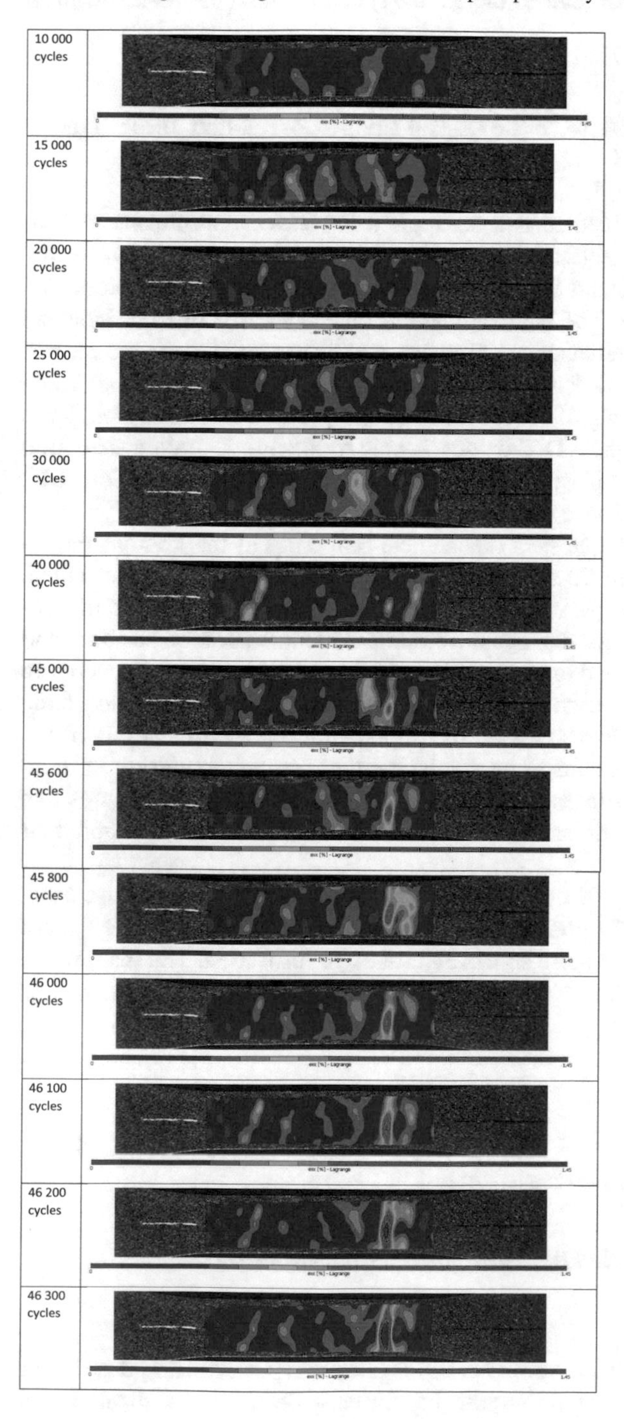

obtained. Subsequent images refer to the coarse specimen with the layer thickness of 40 μm.

6.2.2.5 Final Remarks Related to the Digital Image Correlation Technique Application

Digital Image Correlation method can be applied for tests on specimens with or without geometrical imperfections. It enables capturing damage zones up to the specimen fracture. The method identifies the interaction of strain fields resulting from the notches presence. The isolines created on the surface of multi-notched specimen appeared at the initial stage of tension and disappeared at the moment of damage zones creation and crack appearance. Independently on a type of geometrical imperfections considered, a significant reduction of proportional limit and yield point can be easily indicated. Typical DIC device is able to identify a strain distribution close to the notch for elastic-plastic state. In the case of elastic state, the micro-DIC method using microscope device is more recommended.

Thanks to the DIC technique such mechanical properties as the Young's modulus, yield point, ultimate tensile strength and elongation can be easily determined. More importantly, the technique with some limitations in the accuracy of strain measurement, can be attractive for damage development identification on surfaces and subsurfaces. It can be also used to monitor deformation changes at high-temperature fatigue tests, however, the usage of the induction heater reduces the field of the camera view significantly. Nevertheless, the application of DIC to analysis of the strain distribution changes seems to be a valuable tool for the monitoring of damage development on the surfaces of many responsible elements of structures and machines, including an identification of the places of the cracks initiation and their further propagation.

Due to limited accuracy of the method in terms of damage monitoring, especialy in early stages of the fatigue process development, Electronic Speckle Pattern Interferometry is regarded as more suitable for damage inspection. The main working principles of this technique together with some examples of application will be described in the next section.

6.3 Material Degradation Assessment Supported by Electronic Speckle Pattern Interferometery

6.3.1 Working Principles of Electronic Speckle Pattern Interferometery

Electronic Speckle Pattern Interferometry (ESPI) is the next powerful non-destructive optical method of stress and strain monitoring for early detection, localization and monitoring of damage in materials under monotonic and cyclic loading (Vial-Edwards et al, 2001; Patorski, 2005; Gungor, 2009; Pierron, 2009; Dietrich et al, 2012; Szymczak et al, 2013; DANTEC, 2019). ESPI method allows to monitor a deformation development until the specimen decohesion. A typical measuring system consists of the CCD camera localized in the head of the system and four light sources (Fig. 6.20). A reference speckle pattern formed by the reference beam is also observed by the camera. The typical light source used in ESPI is a 75-mW Nd: YAG laser, emitting a green beam with the wave length of λ=785 nm. It is divided by the beam-splitter into four beams. The image of the specimen due to the reflected laser waves is detected by a CCD (charge-coupled device) sensor of the camera and then transferred to the computer.

The results obtained by means of ESPI are derived from the physical surface deformations. This noncontact and noninvasive technique is used in experimental mechanics to obtain the displacement maps during the loading process. ESPI is based on the application of the wave interference phenomenon. The main purpose of this method is to obtain the fringe patterns which represent waves interference from the coherent light sources illuminating surface of the object. The electromagnetic waves are superimposed and the resulting wave intensity pattern is determined by the phase difference between the waves. The illumination of a rough surface with coherent laser light and subsequent imaging using a CCD camera generates statistical interference patterns, the so-called speckles. These speckles are inherent to the investigated surface. Superimposing a reference light, which is split out of the same laser source, on these speckles results in an interferogram. If the object under test is loaded, and the surface is deformed, the speckle interferogram changes. Comparing an interferogram of the surface before and after loading will result in a fringe pattern, which reveals the displacement of the surface during loading as contour lines of deformation.

As shown in Fig. 6.21, the laser beam is divided into an object beam (blue line) and reference beam (green line). When the object is displaced in the direction normal to the viewing direction, a distance travelled by the object beam changes, and as a consequence, the amplitude of the combined beams becomes altered. The speckle effect is a result of the interference of waves having different amplitudes (Fig. 6.22). This process can be analysed mathematically. Let us consider a pair of

Fig. 6.20 ESPI system: (a) CCD camera, (b) light sources (DANTEC, 2019)

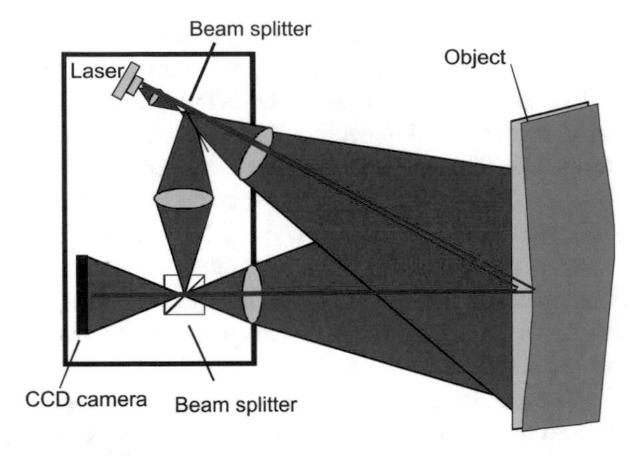

Fig. 6.21 Laser speckle interferometry set-up (DANTEC, 2019)

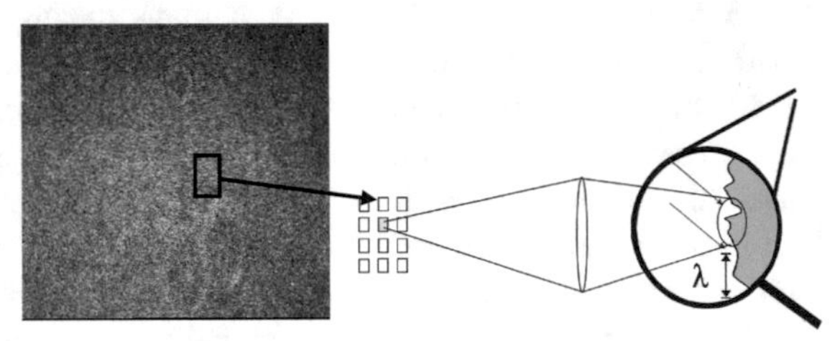

Fig. 6.22 Origin of the speckle (DANTEC, 2019)

simple coherent waves expressed by:

$$\begin{aligned} U_1(x,y) &= A_1 \mathrm{e}^{-\mathrm{i}(\omega t - \varphi_1(x,y))} \\ U_2(x,y) &= A_2 \mathrm{e}^{-\mathrm{i}(\omega t - \varphi_2(x,y))} \end{aligned} \tag{6.5}$$

where (x, y) represent the coordinates of the image plane, A is the magnitude of the displacement, $\varphi(x, y) = \varphi_1(x, y) - \varphi_2(x, y)$ represents the phase difference between the two beams, and ω denotes the angular frequency. The light wave intensity can be found using the following equation:

$$I(x,y) = \int [U_1(x,y) + U_2(x,y)][U_1^*(x,y) + U_2^*(x,y)]dt \tag{6.6}$$

Applying Eqs. (6.5), the intensity of the light wave expressed by Eq. (6.6) can be transformed into the following relationship:

$$\begin{aligned} I(x,y) &= A_1^2 + A_2^2 + 2A_1A_2\cos[\varphi_1(x,y) - \varphi_2(x,y)] \\ &= A_1^2 + A_2^2 + 2A_1A_2\cos[\varphi(x,y)] \end{aligned} \tag{6.7}$$

Thus, the amplitude of the light at any point of the image is the sum of the light amplitude of the reflected wave from the object and that which represents reference beam.

The speckle effect occurs when the rough surface is illuminated by the coherent light, if the surface roughness is greater than the wavelength λ. As the result of the light beam scattering on the object surface, the interference of secondary waves occurs, which leads to the formation of the characteristic speckle pattern.

When the specimen is deformed, the speckle pattern obtained before loading (reference image) is subtracted from the speckle pattern obtained after loading (measurement image), fringes are obtained which represent contours of displacement (see Fig. 6.23). The fringes represent the points of the same displacement. Based on the fringe image, it is not possible to find a direction of displacement.

The intensity distributions $I_1(x, y)$ and $I_2(x, y)$ recorded before and after the object displacement can be expressed by the following equations:

$$I_1(x, y) = A_1^2 + A_2^2 + 2A_1A_2 \cos[\varphi(x, y)] \tag{6.8}$$

$$I_2(x, y) = A_1^2 + A_2^2 + 2A_1A_2 \cos[\varphi(x, y) + \Delta\varphi(x, y)] \tag{6.9}$$

where $\Delta\varphi(x, y)$ is the additional phase change due to the object deformation (Fig. 6.24). Figure 6.24 shows that bright interference bands (containing the maximum intensity values) occur in places where the phase difference is of the even multiple of π. Dark bands (containing the minimum values of intensity) correspond to phase difference of the odd multiple of π. A new interference fringes are obtained by subtracting the signals:

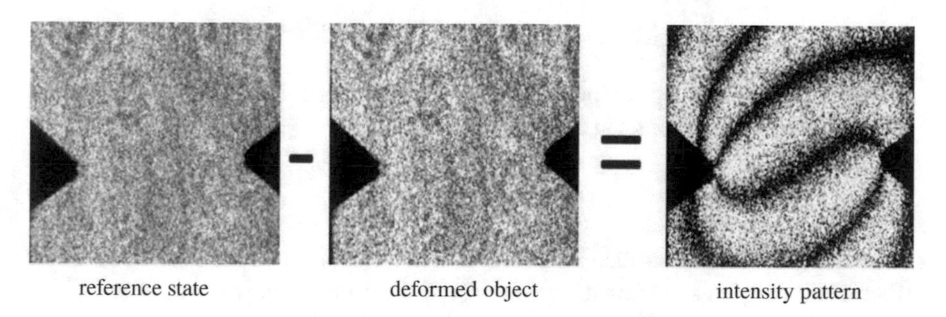

Fig. 6.23 A schema of the fringe patterns determination (DANTEC, 2019)

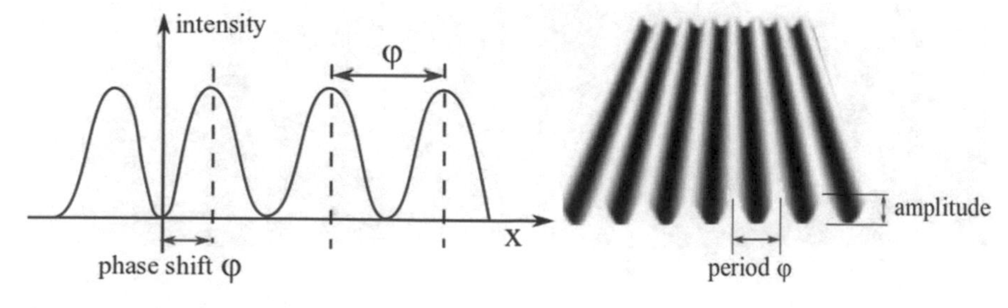

Fig. 6.24 Original sinusoidal signal

$$I(x,y)_{\text{after}} - I(x,y)_{\text{before}} = 2A_1A_2\{\cos[\varphi(x,y) + \Delta\varphi(x,y)] - \cos[\varphi(x,y)]\} \quad (6.10)$$

In the case of adding the signals of the wave intensity one can obtain

$$I(x,y)_{\text{after}} + I(x,y)_{\text{before}} = 2A_1^2 + 2A_2^2 + 2A_1A_2\{\cos[\varphi(x,y) + \Delta\varphi(x,y)] + \cos[\varphi(x,y)]\} \quad (6.11)$$

The intensity of the interference pattern takes the form:

$$I_j(x,y) = A_1^2 + A_2^2 + 2A_1A_2\cos[\varphi(x,y) + \alpha_j] \quad (6.12)$$

where α_j is the amount of phase shifting with $j = 1, 2, 3, \ldots, N$ representing the integer number depending on the phase shifts number introduced. Applying the four-phase shift method $(-3\pi/4, -\pi/4, \pi/4, 3\pi/4)$, the phase of the wavefront computed from the four interferograms is

$$\varphi(x,y) = \arctan\frac{I_4(x,y) - I_2(x,y)}{I_1(x,y) - I_3(x,y)} \quad (6.13)$$

There are three unknown values: $A_1, A_2, \varphi(x,y)$ and α_j is a known optical phase shift. With the intensity maps the optical phase at each pixel can be calculated. Minimum three images with different relative optical path length are acquired. Finally, one can calculate the optical phase for each deformation stage (Fig. 6.25). A difference between the phase maps calculated after and before the deformation is given by

$$\Delta\varphi(x,y) \equiv \varphi(x,y)^{\text{after}} - \varphi(x,y)^{\text{before}} \quad (6.14)$$

On the other hand, the optical path difference (OPD) for the object wavefront relative to the reference wavefront can be determined from:

$$\text{OPD}(x,y) = \frac{\lambda\Delta\varphi(x,y)}{2\pi} \quad (6.15)$$

The OPD is related to surface heights by the multiplicative factor and accounts for illumination angles and viewing which may differ from the normal surface.

Therefore, the phase change

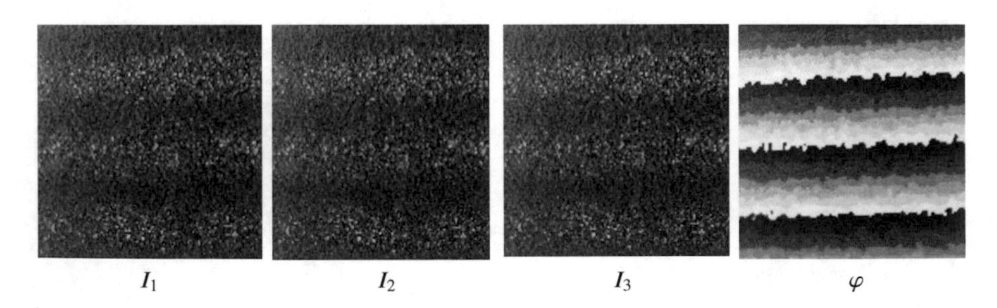

Fig. 6.25 Phase shift algorithm (DANTEC, 2019)

$$\Delta\varphi(x, y) \equiv \varphi(x, y)^{\mathrm{after}} - \varphi(x, y)^{\mathrm{before}}$$

of the wavefront from the object surface before and after deformation is directly associated with the component of displacement $u(x)$ and can be expressed as

$$\Delta\varphi(x, y) = \frac{4\pi}{\lambda} \sin\theta u(x, y) \tag{6.16}$$

where λ is the length of laser wave, and θ is the incident angle between two light beams illuminating the object, and $u(x)$ is the in-plan displacement component.

Finally, one can calculate the optical phase for each deformation stage (Fig. 6.26a). The next point is to unwrap the phase map. A definition of the starting point is required (see Fig. 6.26a). The offset $(n * 2\pi)$ must be taken into account. The displacement is encoded as levels of gray that represent the intensity at a given pixel $I(x, y)$ of the scalar field. Finally, the continuous phase map is determined, Fig. 6.26b.

In the next step the strain is calculated. Determination of the fringe pattern gradient is shown in Fig. 6.27. Taking into account Fig. 6.27, the following relationship can be formulated:

$$\tan\theta_u = \frac{\dfrac{\partial u}{\partial y}}{\dfrac{\partial u}{\partial x}} \tag{6.17}$$

Applying the relationship between the derivative of displacement components and the fringe orientation of the object before and after deformation, the strain components can be found according to the following equations

$$\frac{\partial u}{\partial x} = \varepsilon_{xx}, \quad \frac{\partial u}{\partial y} = \gamma_{xy}, \quad \frac{\partial v}{\partial x} = \gamma_{yx}, \quad \frac{\partial v}{\partial y} = \varepsilon_{yy} \tag{6.18}$$

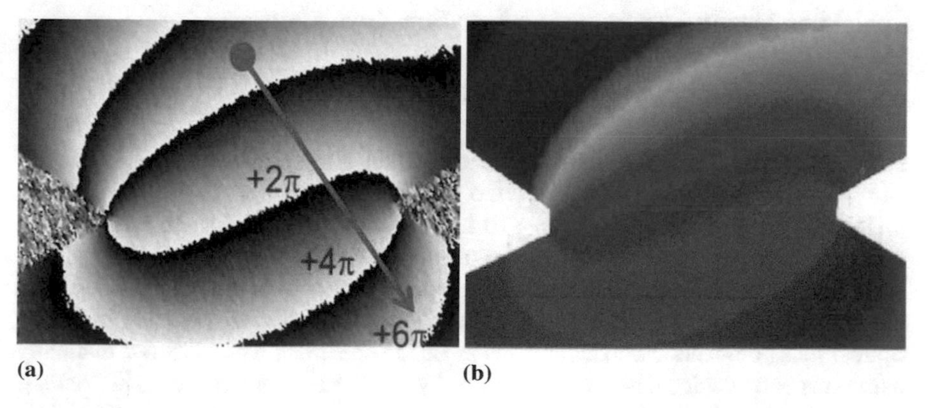

Fig. 6.26 Elaboration of ESPI measurements: (a) Phase map, (b) Continuous phase map (DANTEC, 2019)

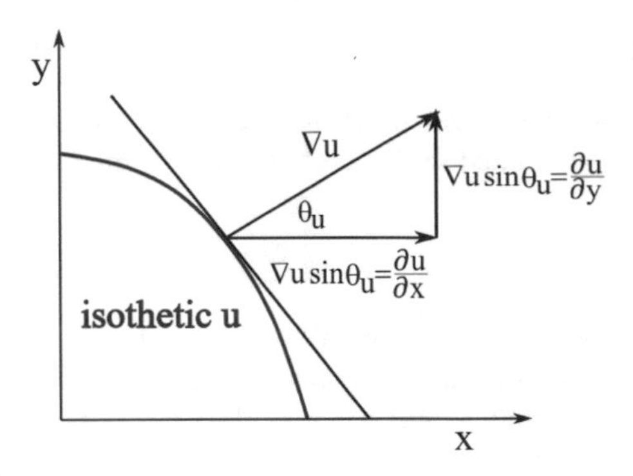

Fig. 6.27 Gradient vector and the partial derivatives with respect to the coordinate axes

The principal strain components are defined by the formulas

$$\begin{aligned} \varepsilon_{h_1} &= \frac{\varepsilon_{t_1} + \varepsilon_{t_2}}{2} + \sqrt{\left(\frac{\varepsilon_{t_1} - \varepsilon_{t_2}}{2}\right)^2 + \left(\frac{\gamma_{t_1 t_2}}{2}\right)^2}, \\ \varepsilon_{h_2} &= \frac{\varepsilon_{t_1} + \varepsilon_{t_2}}{2} - \sqrt{\left(\frac{\varepsilon_{t_1} - \varepsilon_{t_2}}{2}\right)^2 + \left(\frac{\gamma_{t_1 t_2}}{2}\right)^2} \end{aligned} \tag{6.19}$$

ESPI is implemented by producing interference between an optical wavefront scattered from the object and the fixed reference wave, giving as a consequence the displacement components for each point in the image of the object. The use of this method during fatigue testing enables the location of the greatest concentration of stress-induced defects to be determined, and more importantly it provides necessary data for damage initiation predictions with relatively high accuracy. This issue will be wider presented in the next subsection.

6.3.2 Representative Applications of Electronic Speckle Pattern Interferometry in the Laboratory Investigations

6.3.2.1 The Results for the Nickel Alloy

Fatigue investigations were carried out on nickel alloy (C – 0.09%, Cr – 8.8%, Mn – 0.1%, Si– 0.25%, W– 9.7%, Co – 9.5%, Al – 5.7%, Ta+Ti+HF– 5.5%) under force control using the MTS-810 hydraulic testing machine. In each test the maximum cyclic stress range and stress amplitude were equal to 600 MPa and 300 MPa, respectively. Both these parameters were lower than the yield point of the material in question. In order to eliminate vibrations of the testing machine during optical measurements the loading process of the specimen was executed manually using a special device designed originally by the IPPT PAN workers. The loading programme is presented in Fig. 6.28.

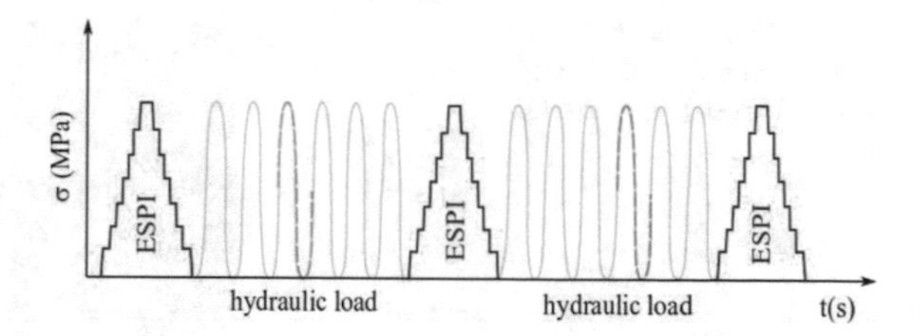

Fig. 6.28 Scheme of loading during fatigue test

As it is shown, the first cycle was conducted manually, and subsequently, a block of cycles under the frequency of 10 Hz was carried out using the testing machine (Kopeć et al, 2012). The process of cyclic loading was interrupted several times in order to perform displacement measurements by means of the ESPI camera. The experimental programme provided displacement measurements at the beginning of test and after 20000, 40000 and 50000 cycles. The number of cycles to failure was $N_f = 54315$.

ESPI observations carrying out at various stages of the fatigue degradation represent a status and dynamics of the damage development. They enable a determination of the areas of the greatest stress concentrations and reflect a local character of the fatigue damage initiation.

The field strain distributions along the Y axis corresponding to the acting stress direction in the specimen are presented in Fig. 6.29. The figure shows strain distributions for different stages of the fatigue process, i.e. after: (a) first cycle; and (b) 50000 cycles. It has to be mentioned, that all these maps were obtained for the same scale in order to enable a direct comparison of the results achieved. As it is seen, the method enables identification of places where damage initiates. Figure 6.30 demonstrates a location of decohesion on the specimen gauge length, that well agrees with the largest displacement concentration occurring on the phase maps captured by means of the ESPI measurements.

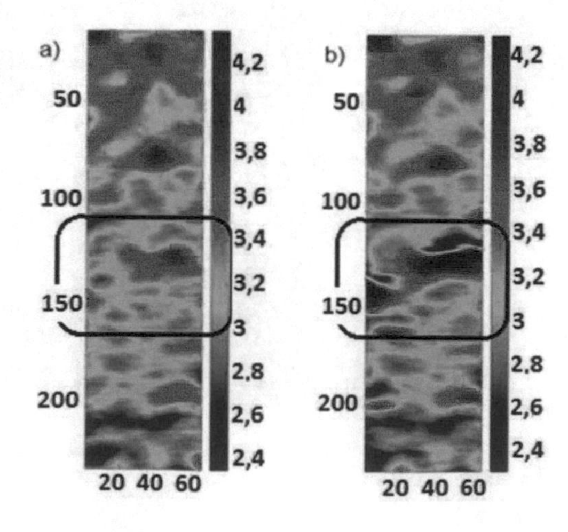

Fig. 6.29 The field strain distribution along Y axis corresponding to the acting stress direction. Measurements performed: (a) in the first loading cycle; (b) after 50000 cycles. Scale is matched to the extreme strain values in the first measurement

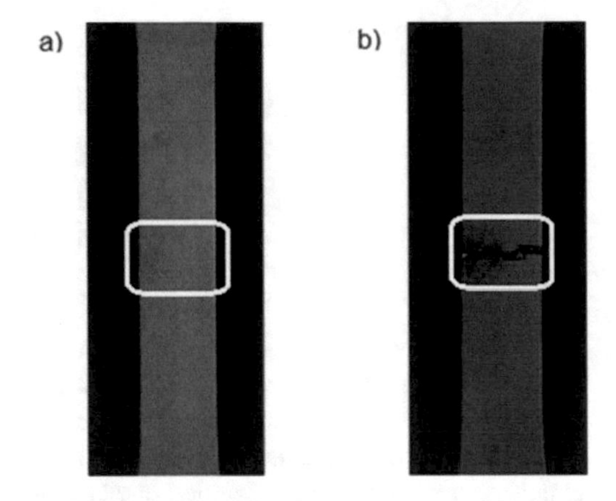

Fig. 6.30 Fatigue specimen photo: (a) before the test; (b) after the test ($N_f = 54315$ cycle)

In Fig. 6.31 another example of the strain distribution map obtained using ESPI for nickel alloy is presented for increasing number of loading cycles. The lateral profiles of maximal cross-sectional strain distribution for increasing cycles number are presented in Fig. 6.32. They enable identification of the highest values of strain on the strain distribution maps, and therefore, provide info for the indication of potential places of damage initiation.

6.3.2.2 The Results for the P91 Steel

The specimens used in this study were manufactured using X10CrMoVNb9-1 (P91) polycrystalline steel. This is a low carbon, creep-resistant steel, typically used for

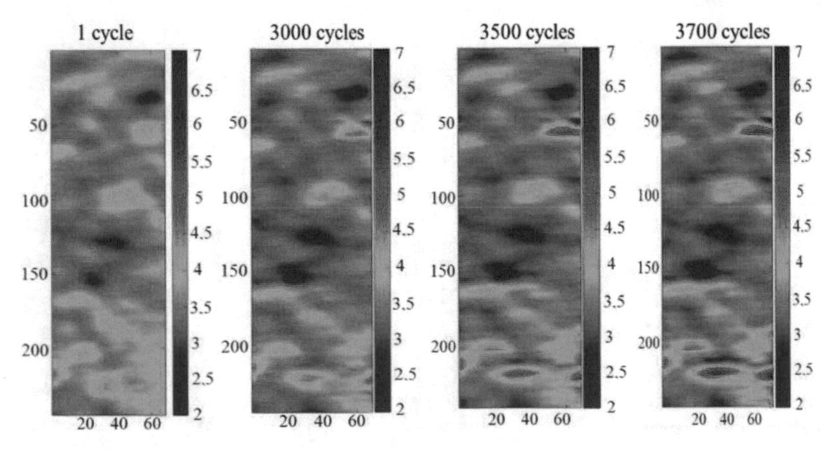

Fig. 6.31 Strain distribution map on the plane specimen surface using ESPI for different stages of the fatigue process

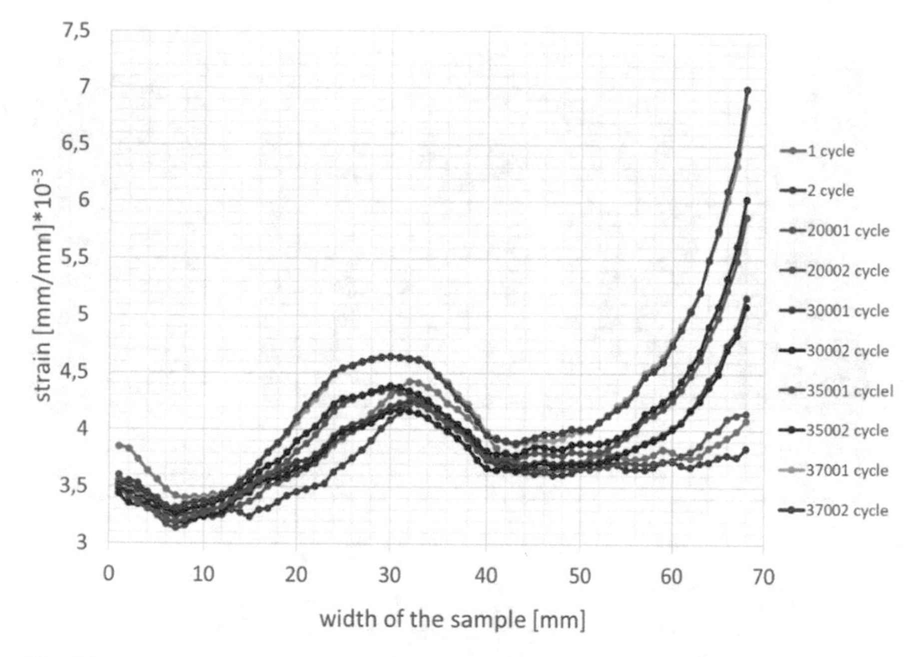

Fig. 6.32 The lateral profiles of the maximal cross-sectional strain evolution

tubes, plates and structural components in the power plant industry. The content of alloying elements in P91 is given in Table 6.2.

Table 6.2 Chemical composition of the P91 steel (in wt. %)

Element	C	Mn	Cr	Mo	V	Ni	Cu	Si	S
P91	0,2÷0,5	0,3÷0,6	8÷9,5	0,85÷ 1,1	0,18÷0,25	< 0,4	< 0,3	0,08÷0,12	< 0,01

The purpose of tests on P91 steel was to provide experimental data of the evolution of micro-strain regime during high cycle fatigue crack initiation. The whole fatigue process was divided on the blocks of cycles and carried out using the hydraulic servocontrolled testing machine. The process of cyclic loading was interrupted several times after selected numbers of cycles in order to perform displacement measurements by means of the ESPI camera. Strain distribution maps at the maximum load applied on the plane specimen surface using ESPI are presented for increasing number of the loading cycles in Figs. 6.33, 6.34 and 6.35. Three types of specimens were tested: specimen with relatively rough surface, electropolished specimen and specimen with the hole. The different stages of the fatigue process, from the beginning up to the moment of crack initiation and subsequent propagation are presented.

During the process of cyclic loading the material deforms heterogeneously and numerous strain concentration spots are visible in the case of rough specimen, Fig.

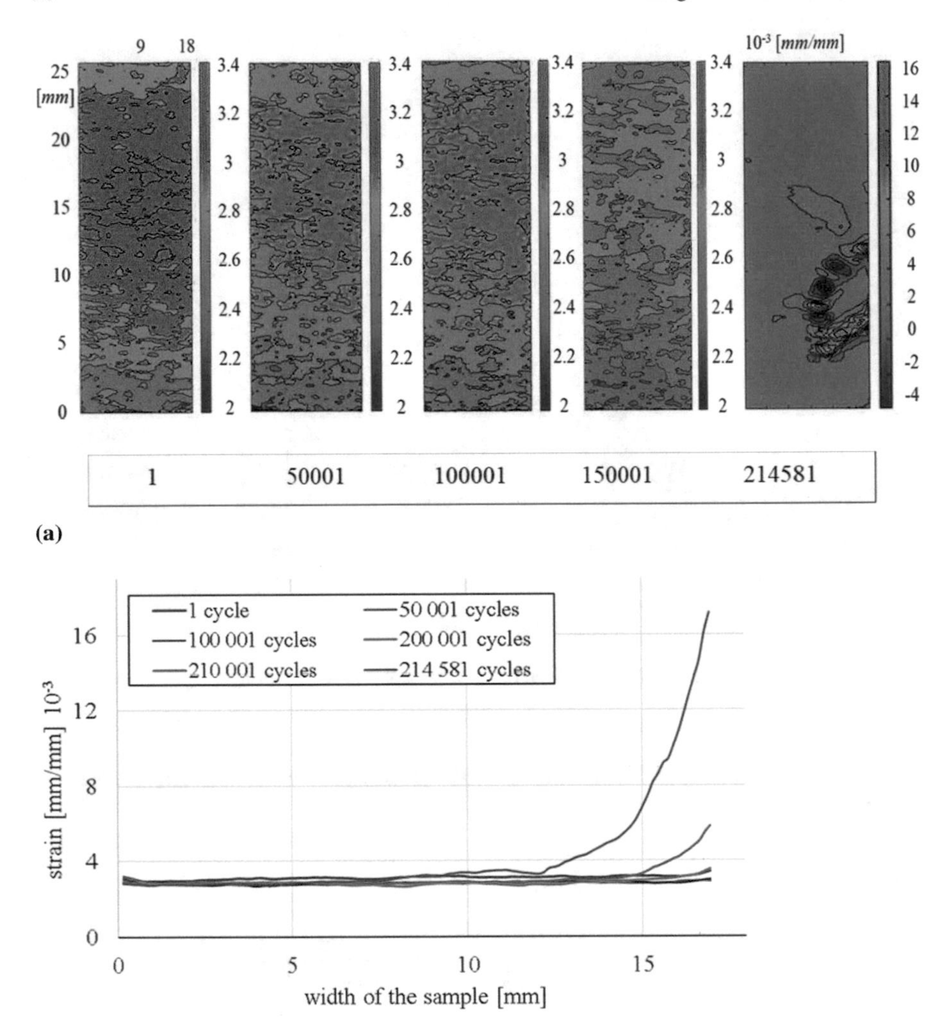

Fig. 6.33 The lateral profiles of the maximal cross-sectional strain evolution

6.33a. The lateral profiles of the maximal cross-sectional strain distribution for increasing cycles number for rough specimen on the surface are presented in Fig. 6.33a. It is expected that the microstructure of the material plays an important role in the strain localization. In order to reduce roughness and remove some modification of the material properties resulting from electro-machining (local heating-cooling), the surface of specimen was electro-polished in the vicinity of cracks generated in the fatigue process.

The strain distribution maps are presented in Fig. 6.34a. The lateral profiles of the maximal cross-sectional strain distribution for increasing cycles number for

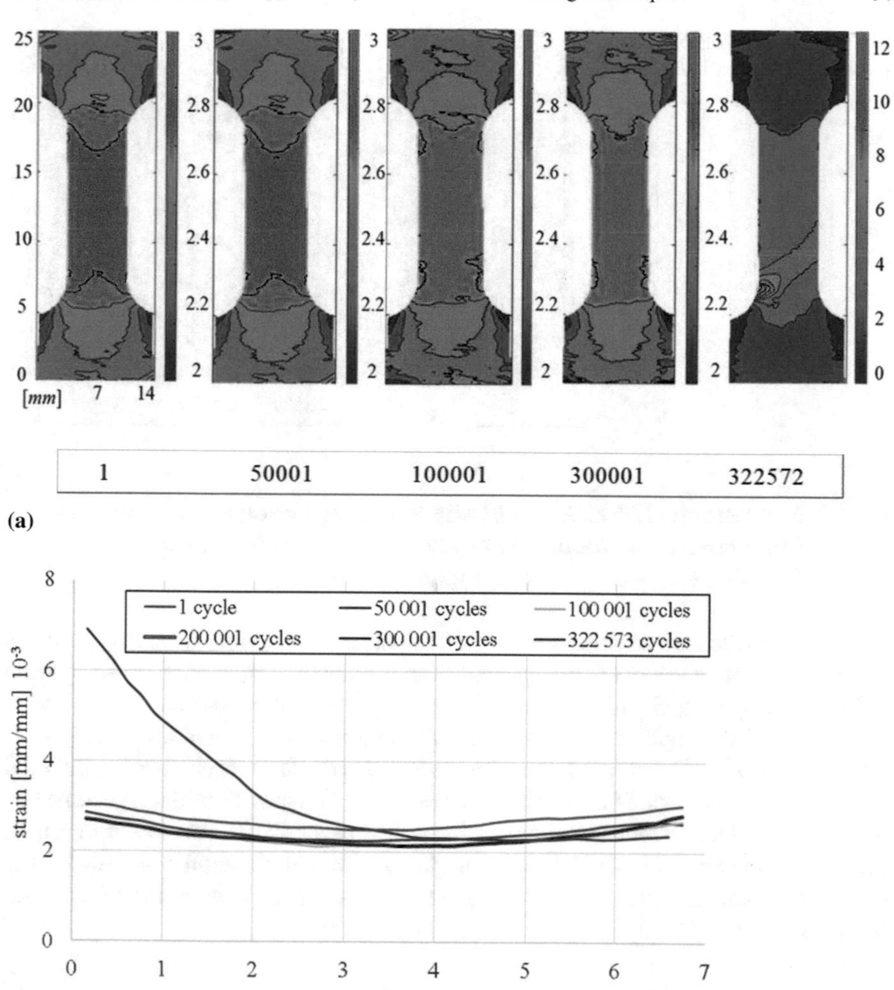

Fig. 6.34 The lateral profiles of the maximal cross-sectional strain evolution

electropolished specimen are presented in Fig. 6.34b. As it is seen from the profiles, Figs. 6.33b and 6.34b, the places of damage initiation are simple for identification. They are well represented by the significant increase of strain.

In the case of specimen with the hole, in the first cycle the strain accumulation spot is visible near the hole, (Fig. 6.35, 1 cycle). In the next cycles the material is strengthened (Fig. 6.35, 2 cycles), but the local zones of strain accumulation in the vicinity of the hole can be still observed, however, they have slightly lower level in comparison to the first cycle. The crack localisation on the left side of the hole is presented in Fig. 6.35 after 78057 cycles.

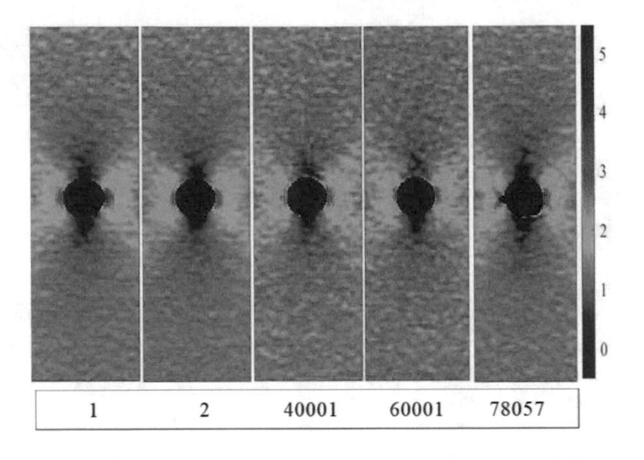

Fig. 6.35 Strain distribution maps - specimen with the hole

6.3.2.3 Mathematical Modelling of Fatigue Damage Evolution, Numerical Implementation Supported by Electronic Speckle Pattern Interferometry Results (Ustrzycka et al, 2017)

In this section the mathematical description of fatigue crack initiation and evolution is formulated. The problem of damage evolution for metals subjected to cyclic loading inducing fatigue crack initiation and its propagation within the elastic regime is considered. The condition of damage accumulation is formulated after Mróz et al (2005). It is assumed that, when the critical stress condition is reached on the material plane, the damage zone Ω is generated. Afterwards, the growth of damage zone can be described. The mathematical model is applied to study damage evolution under cyclic tension and the predictions are compared with experimental data. The profile of normal strain $\varepsilon(x)$ and stress $\sigma(x)$ along the damage zone Ω is expressed as a sum of mean $(\bar{\varepsilon}, \bar{\sigma})$ and fluctuation $(\tilde{\varepsilon}(x), \tilde{\sigma}(x))$ components

$$\varepsilon(x) = \bar{\varepsilon} + \tilde{\varepsilon}(x), \qquad \sigma(x) = \bar{\sigma} + \tilde{\sigma}(x) \tag{6.20}$$

The material is assumed to be linearly elastic, but exhibiting a damage process at strain concentration zones. In order to illustrate the problem (Fig. 6.36), a potential damage zone Ω is selected with the largest strain and stress fluctuations.

The damage evolution rule (6.21) was originally formulated by Mróz et al (2005) for brittle materials

$$dD = A \left(\frac{\sigma - \sigma_0^*}{\sigma_c - \sigma_0} \right)^n \frac{d\sigma}{\sigma_c^* - \sigma_0^*} \tag{6.21}$$

where A and n denote material parameters, σ_0 is the damage initiation threshold values, σ_c denotes the failure stress in tension for the damaged material, σ_0^* and σ_c^* are the threshold values for the undamaged material and D denotes the scalar measure of damage $(0 \leq D \leq 1)$.

The stress value σ increases but the values of σ_0 and σ_c decrease. Both σ_0 and σ_c depend on the damage state according to the formula

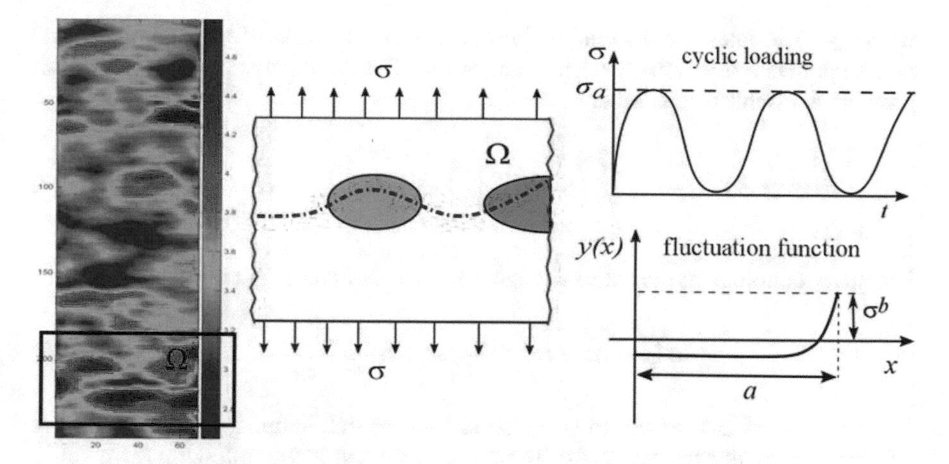

Fig. 6.36 Damage zone Ω with the major stress fluctuation

$$\sigma_c - \sigma_0 = (\sigma_c^* - \sigma_0^*)(1 - D)^p \tag{6.22}$$

where p is a material parameter. The process of cyclic loading is described by the time variation of stress in the following form

$$\sigma(t) = \frac{1}{2}\sigma_a[1 + \sin(\omega t)], \qquad \omega = \frac{2\pi}{T}, \tag{6.23}$$

where σ_a is the stress amplitude. Substitution of σ from Eq. (6.23) into Eq. (6.21) and integration lead to the equation of damage increase in a single cycle that can be expressed as follows

$$\int_0^{\widetilde{D}} (1 - D)^{np} dD = \frac{A\omega\sigma_a}{2(\sigma_c^* - \sigma_0^*)^{n+1}} \int_0^T \left\{\frac{1}{2}\sigma_a[1 + \sin(\omega t)] - \sigma_0^*\right\}^n \cos(\omega t)dt \tag{6.24}$$

Finally, damage evolution law for single cycle takes the form:

$$\widetilde{D} = 1 - \left\{1 - \frac{A\omega\sigma_a}{2(\sigma_c^* - \sigma_0^*)^{n+1}} \left[\left(\frac{1}{2}\sigma_a[1 + \sin(\omega t)] - \sigma_0^*\right)^{n+1} - \left(\frac{1}{2}\sigma_a - \sigma_0^*\right)^{n+1}\right]\right\}^{\frac{1}{pn+1}} \tag{6.25}$$

The free edges of the sample due to surface irregularities act as a kind of stress concentrators. The influence of edge defects on the damage evolution and crack propagation is significant.

In order to account for the edge effect, the stress fluctuation function (see Fig. 6.36) is introduced and the total stress expressed as follows

$$\sigma(x) = \bar{\sigma} + \tilde{\sigma}(x) = \bar{\sigma} + \bar{\sigma}y(x) = \bar{\sigma}\left(1 + \alpha + \beta\left|\frac{x}{a}\right|^m\right) \tag{6.26}$$

where $y(x)$ denotes the fluctuation function, a is the width of the sample, β and m are the material parameters. The integration of stress fluctuation on $[0, a]$ makes possible to establish α parameter

$$\int_0^a \tilde{\sigma}(x)dx = 0 \quad \rightarrow \quad \int_0^a \left(\alpha + \beta\left|\frac{x}{a}\right|^m\right) dx = 0 \quad \rightarrow \quad \alpha = -\frac{\beta}{m+1} \tag{6.27}$$

Boundary condition at the external edge allows to designate β parameter

$$\tilde{\sigma}(x = a) = \sigma^b \quad \rightarrow \quad \beta = \frac{\sigma^b}{\tilde{\sigma}}\frac{m+1}{m} \tag{6.28}$$

The value of σ^b is assumed to correspond to measured boundary fluctuation, here $\sigma^b = 1.1\tilde{\sigma}$. Finally, the stress distribution in expressed in the following form

$$\sigma(x,t) = \bar{\sigma}\left(1 + \alpha + \beta\left|\frac{x}{a}\right|^m\right)\left[\frac{1}{2} + \frac{1}{2}\sin(\omega t)\right] \tag{6.29}$$

According to the mathematical description, the numerical analysis of damage evolution (6.21) under mechanical loads (6.29) in elastic-plastic solids has been made. The evolution of damage during the increasing number of cycles is shown in Fig. 6.37. The macro-crack initiation occurs at the critical value of damage $D_c = 0.3$.

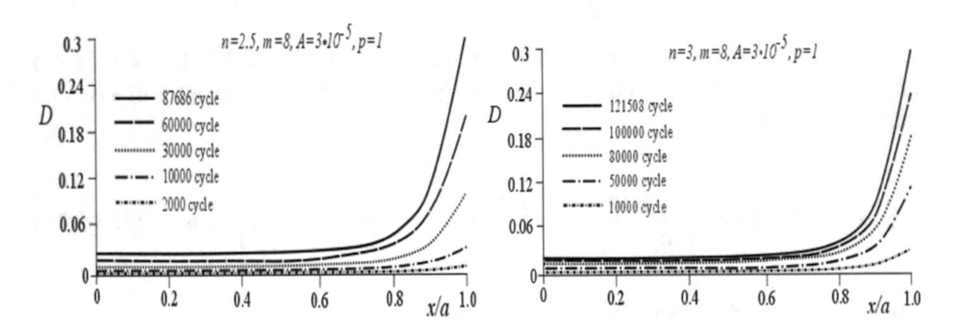

Fig. 6.37 Damage evolution related to the number of cycles for different values of the parameter n

6.4 Alternative Non-destructive Testing Techniques for Damage Identification

6.4.1 Introductory Remarks

In the previous sections the optical nondestructive techniques were presented as the powerful tools for fatigue damage identification. It has to be mentioned however, that there are many other nondestructive testing techniques commonly used for damage assessments (Narayan and Green Jr., 1975; Sablik and Augustyniak, 1999; Marténez-Ona and Pérez, 2000; Ogi et al, 2000; Fel et al, 2001). There are also many destructive methods supporting this issue (Hayhurst, 1972; Trąmpczyński and Kowalewski, 1986; Kowalewski, 2005). Having the parameters of destructive and nondestructive methods for damage development evaluation, it is instructive to analyze their variation in order to find possible correlations. This is because of the fact that typical destructive investigations, like creep, fatigue or standard tensile tests, give the macroscopic parameters characterizing the lifetime, strain rate, yield point, ultimate tensile stress, ductility, etc. without any information concerning microstructural damage development and material microstructure variation. On the other hand, nondestructive methods provide information about damage at a particular time of the entire working period of an element, however, without sufficient information about the microstructure and how it varies with time. Therefore, it seems reasonable to plan future damage development investigations in the form of interdisciplinary tests connecting results achieved using destructive and nondestructive methods with microscopic observations in order to find mutual correlations between their parameters (Dietrich and Kowalewski, 1997; Makowska, 2014).

To assess damage using destructive method, the specimens after different amounts of prestraining are usually stretched up to failure. The selected tensile parameters can be determined afterwards, and their variations can be used for identification of damage development. Ultrasonic and magnetic investigations are often selected as the nondestructive methods for damage development evaluation. In the case of ultrasonic method, the acoustic birefringence coefficient is often used to identify damage development in the steels tested. Two magnetic techniques for the nondestructive testing are especially suitable, i.e., measurement of the Barkhausen effect (HBE) and the magneto-acoustic emission (MAE). Both effects are due to an abrupt movement of the magnetic domain walls depicted from microstructural defects when the specimen is magnetized. The laboratory test specimens are magnetized by a solenoid and the magnetic flux generated in the specimen is closed by a C-core shaped yoke. The magnetizing current (delivered by a current source) has typicaly a triangular waveform and frequency of order 0.1 Hz. Its intensity is proportional to the voltage UG. Two sensors can be used: (a) a pickup coil (PC) and (b) an acoustic emission transducer (AET). A voltage signal induced in the PC is used for the magnetic hysteresis loop $B(H)$ evaluation (low frequency component) as well as for the HBE analysis (high frequency component). The intensity of the HBE is given by the rms (root mean square) voltage U_b envelopes. The maximal values (U_{bpp}) of U_b for one period of magnetization can be then compared. An analogous analysis can be performed for the MAE voltage signal from the AET. In this case, the maximal values (U_{app}) of the U_a voltage envelopes are compared. The magnetic coercivity H_c, evaluated from the $B(H)$ hysteresis loop plots, can also be taken into account.

6.4.2 Examples of the Results from Research Programs Executed for Damage Analysis

6.4.2.1 Magnetic Techniques Combined with Destructive Methods

This section describes an application of non-destructive magnetic techniques supporting damage evaluation due to creep or plastic flow on the basis of the selected damage sensitive parameters coming from tensile tests and magnetic investigations carried out on the P91 steel specimens. The low carbon, creep-resistant P91 steel is typically used for tubes, plates and structural components in the power engineering. Two different types of deformation processes were carried out. First, a series of the P91 steel specimens was subjected to creep, and second one to plastic deformation in order to achieve the material with an increasing strain level up to 10%. Subsequently, non-destructive and destructive tensile tests were performed. Magnetic methods based on measurements of magnetoacoustic emission and magnetic hysteresis loop changes were applied. The static tensile tests were carried out in order to evaluate the mechanical parameters variations. It is shown that some relationships between the selected parameters coming from the non-destructive and destructive tests may be formulated.

The magnetoacoustic emission (MAE) technique is based on the analysis of acoustic signals generated in the bulk of the material subjected to alternating magnetic fields (Makowska and Kowalewski, 2015). MAE is generated during the movement of non-180° domain walls (in the case of steels they are 90° domain walls) as a result of local deformations (volume changes) induced by the local change of magnetisation in a material having the non-zero magnetostriction (Buttle et al, 1986). Domain walls are pinned temporarily by microstructural barriers to disable their motion, and then they are released abruptly with an increase of the magnetic field. Grain boundaries, precipitates, dislocation tangles (Jiles, 1998) and voids (Blaow et al, 2007) are microstructural barriers hindering domain walls movement. It was also suggested that irreversible rotation of the domain through angles other than 180° can also contribute to the MAE signal (O'Sullivan et al, 2004). Magnetoacoustic emission accompanies magnetic Barkhausen effect (MBE) that is produced in steels by movement of both 180° and 90° domain walls. Movement of the 180° domain walls (between anti-parallel domains) does not contribute to the generation of elastic waves. Stresses are not generated as the result of 180° domain walls movements or rotations, since the strain along a particular axis is independent on the direction of the magnetic moments if they act along the same axis (Buttle et al, 1986). As a consequence, the movement of 180° domain walls does not affect magnetostrictive strain (Buttle et al, 1986). An advantage of the MAE method over the MBE method results from the depth of measurements. In the case of MAE it is dependent only on the ability to magnetise the investigated material and easily reaches 10 mm (or even more in special cases) while for MBE it is only 0–1mm due to attenuation of the generated electromagnetic waves (in the kHz range) by eddy currents (Blaow et al, 2007).

Magnetoacoustic emission intensity, coercivity, saturation induction were measured on the plain specimens of the P91 steel (having a cross section 5 mm × 7 mm and gauge length of 40 mm). The specimens were earlier subjected either to creep (T = 773 K, σ = 290 MPa) or plastic deformation (T = 298 K, V = 1mm/min). The loading process of each specimen was carried out to achieve the different strain level. The creep process was interrupted to obtain various deformation levels i.e.: 0.85%, 1.85%, 3.15%, 4.60%, 5.90%, 7.90% and 9.30%, whereas the levels of plastic deformation were as follow: 2.00%, 3.00%, 4.50%, 5.50%, 7.50%, 9.00%, 10.50%.

Envelopes of the magnetoacoustic emission signal were calculated according to the equation:

$$U_a(h) = \sqrt{\frac{\int_0^\tau U_{ta1}^2(t)dt}{\tau}} \tag{6.30}$$

where: $U_a(h)$ is the root mean square voltage from the acoustic wave sensor calculated over an interval τ during which the average magnetic field $U_{ta1}(t)$ is the output of the sensor. The integral over a half cycle of the magnetic field was calculated as

$$\mathrm{int}(U_a) = \int_{H_{min}}^{H_{max}} U_{sa}(h)dh \tag{6.31}$$

where

$$U_{sa}(h) = \sqrt{U_a{}^2(h) - U_{ta}^2} \tag{6.32}$$

and U_{ta} is the root mean square of the background noise voltage. Hysteresis loop changes were characterised by means of coercivity H_C and saturation induction B_S. The magnetic parameters were normalized with respect to their values for the non-deformed specimen ($\mathrm{Int}(U_a)_{norm}, H_{c\,norm}, B_{s\,norm}$).

Example envelopes of magnetoacoustic emission are presented in Fig. 6.38. The MAE envelope of the non-deformed specimen reveals a broad maximum with two peaks (Fig. 6.38a). According to Kwan et al (1984) the first peak on the MAE envelope is mainly due to the creation and the second one to the annihilation of magnetic domains with high contribution of displacement of non-180° domain walls. The changes in the height and width of the peaks in Fig 6.38b, c indicate that magnetoacoustic emission is sensitive to material damage and depends on the type of deformation. The two-peak broad maximum observed in the non-deformed specimen transforms to a single maximum for the specimen strained up to 10.5% in the tensile test (Fig. 6.38b) as well as for the specimen strained up to 9.3% in the creep test (Fig. 6.38c). It can be also noted that the single maximum for the specimen after plastic

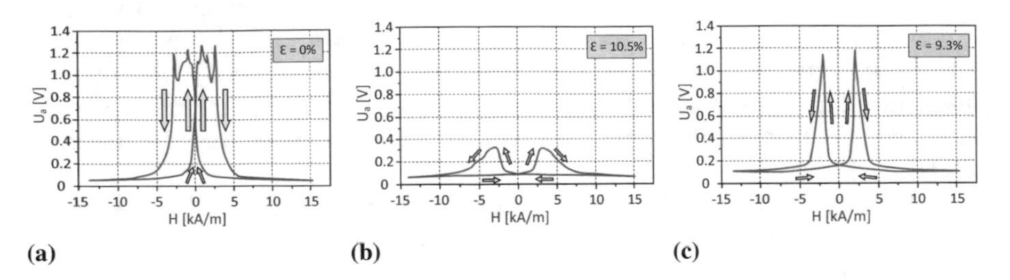

Fig. 6.38 The magnetoacoustic emission rms envelopes for specimens of the P91 steel: a) non-deformed, b) after plastic deformation up to 10.5%, c) after creep up to 9.3% (Makowska et al, 2014)

deformation is broader and lower than for the specimen after creep (Fig. 6.38b, c), which will be discussed later. It can also be seen that the maxima of both envelopes of strained specimens (by creep and plastic deformation) occur at a higher magnetizing field than the maximum of the non-deformed specimen.

Possible explanation is related to the necessity of usage a higher magnetic field to be able to move domain walls in the material with a higher density of dislocations. Transformation of the peak shape for all deformed specimen occurs. The integrals of half-period voltage signals from the magnetoacoustic emission rms envelopes for each specimen were calculated. They are presented in Fig. 6.39.

The integral of the half-period voltage signal of the MAE $\mathrm{Int}(U_a)_{norm}$ decreases with the increase of strain level for both plastic and creep deformations, but the dynamics of these processes are different—lower values of this parameter (for deformation values greater than ε = 2%) were obtained for specimens after plastic deformation (Fig. 6.39). The integral of the MAE is almost insensitive to creep in the range of strain between 0.85% and 9.3%. The decrease of both parameters may be explained on the basis of previous knowledge provided by other researchers (Augustyniak, 2003; Augustyniak et al, 2008). The plastic deformation produces defects inside the martensite plates in the form of dislocation tangles that decrease signif-

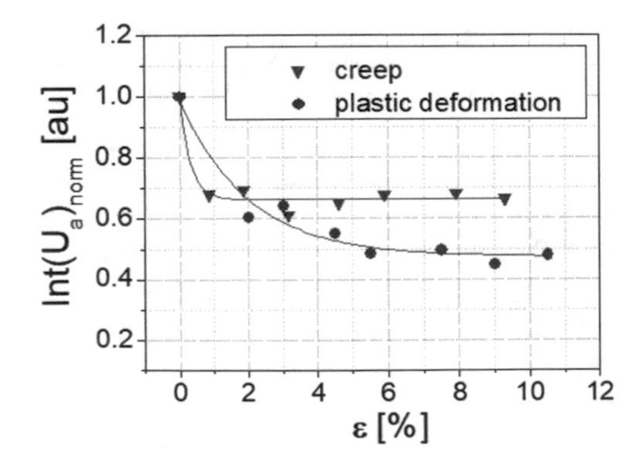

Fig. 6.39 Integral of half-period voltage signal of magnetoacoustic emission versus pre-strain for the P91 steel

icantly the mobility of 'non-180°' domain walls (Augustyniak, 2003). An increase in dislocations density reduces the mean free path of the domain walls displacement and increases their pinning force (Augustyniak et al, 2008). According to the Granato-Lucke theory, it is believed that the ends of the dislocations lines are fastened at points of strong and weak fixation. The points of strong fixation are the nodes of the dislocation network, whereas the points of weak fixation are impurity atoms (Baldev et al, 2001). Under applied stresses the dislocations segments bend between the points of weak fixation. In the case, when the stress exceeds a defined value, the dislocation segment breaks away from the points of weak fixation (Baldev et al, 2001). As a result, due to the dislocation structure modification, movement of the domain walls becomes less effective and the MAE intensity signal and its parameters Int(U_a) decrease monotonically (Blaow et al, 2007).

Usually, in the case of creep of metallic materials, two main processes are dominant: strain hardening and thermal softening (also called as recovery) (Bailey, 1935; Zakharov et al, 2013). The same situation takes place in the P91 steel. The material recovery occurs by dislocation cross slip and dislocation climb. As a result of recovery and relatively high level of acting stress equal to 290 MPa, polygonization of the material occurs thanks to the dislocation climb mechanism. The second important process involved during creep of the P91 steel is the hardening caused by introduction of high density dislocation tangles to the material structure. Therefore, the accelerated creep is a mixed process consisting of creep and plastic components of deformation due to the high level of stress applied. Similarly, for plastic deformation, the movement of the domain walls is impeded due to the significant amount of defects introduced to the material. However, a lesser decrease in the $\mathrm{Int(U_a)_{norm}}$ values for specimens after accelerated creep than for specimens after tensile tests was observed due the applied temperature (773 K).

The results of mechanical tests conducted on specimens with prior deformation were used to determine the yield point and ultimate tensile stress variations. They demonstrate the softening effect in the P91 steel after creep, whereas after plastic deformation a hardening effect can be observed.

The main aim of the research program was to find relationships between damage sensitive parameters of tensile tests and those determined from magnetic investigations. Representative results are shown in Figs. 6.40 - 6.42. They show the relationships between the ultimate tensile stress and selected magnetic parameters.

Similarly to the results obtained for the yield stress in the case of plastic deformation, the exponential relationships between the ultimate tensile stress and magnetoacoustic emission integral can be observed in the strain levels considered (0–9%) here, Fig. 6.40. It has to be noted however, that for the steel after creep the mutual relationship between these parameters cannot be expressed as a function, since the points representing increasing level of deformation are not located in the orderly manner. The same types of relationships were found between the ultimate tensile stress and coercivity, Fig. 6.41. Contrary to these results, variations of the saturation induction $B_{s\,\mathrm{norm}}$ allow an estimation of the ultimate tensile stress of the P91 steel either after creep (in the whole range) or plastic deformation (range 0 – 9%), Fig.

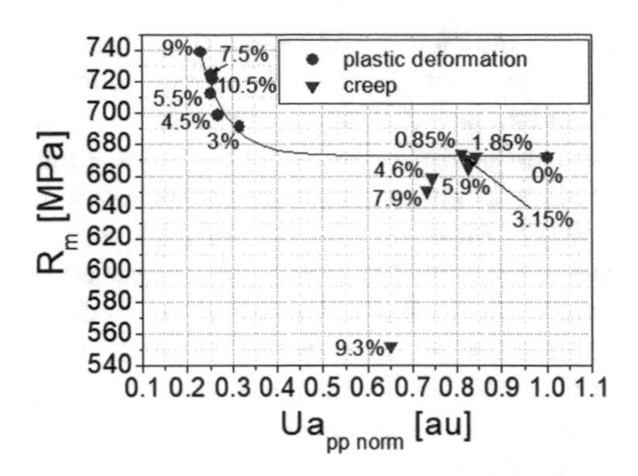

Fig. 6.40 Variation of ultimate tensile stress of the P91 steel versus amplitude of magnetoacoustic emission $Ua_{\mathrm{pp\,norm}}$

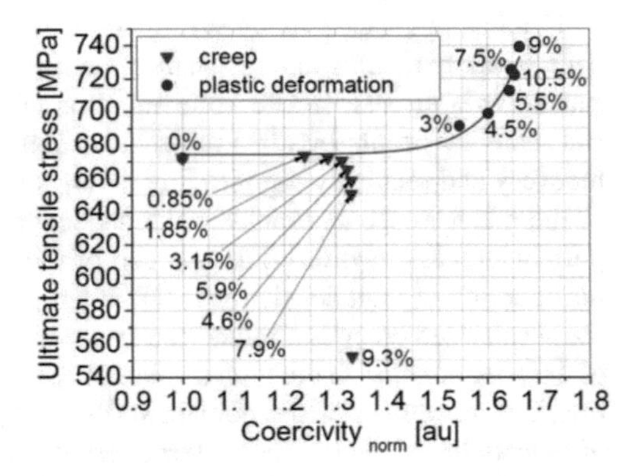

Fig. 6.41 Variation of ultimate tensile stress of the P91 steel versus coercivity $H_{\mathrm{c\,norm}}$

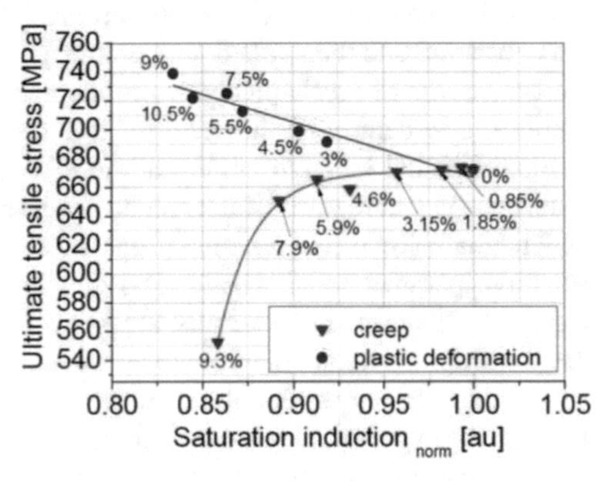

Fig. 6.42 Variation of ultimate tensile stress of the P91 steel versus saturation induction $B_{\mathrm{s\,norm}}$

6.42. The relation $R_{\mathrm{m}} = f(B_{\mathrm{s\,norm}})$ for the material after plastic deformation can be described by the linear function, whereas that after creep, by the exponential one.

6.4.2.2 Ultrasonic Techniques Combined with Destructive Tests

It has been found that various ultrasonic techniques can be applied to evaluate the quality of the material in the as-received state and after deformation history induced by creep, fatigue or plastic flow. Mutual relationships between parameters of ultrasonic waves and those characterising creep were observed in the past (Frost and Ashby, 1982; Brathe, 1978). Variations of nonlinear acoustic parameters were investigated during creep (Ohtani et al, 2006). In late creep stages, when numerous voids were created in the bulk steel specimens, an ultrasonic technique based on acoustic birefringence was used to detect material damage (Kim et al, 2009).

Measurement of acoustic birefringence and evaluation of magnetic parameters seem to be able to provide more comprehensive material degradation data than the replica technique and destructive tests. Acoustic birefringence can be measured using ultrasonic echo technique, for elements accessible from one side only (like pipes for example). Its value is proportional to the difference in the round-trip travel time of ultrasonic pulses polarized in the direction parallel and perpendicular to the loading direction and, at the same time, perpendicular to the loading direction of the specimen. It can be calculated as (Szelązek et al, 2009; Schneider, 1995):

$$B = 2\frac{V_p - V_l}{V_l + V_p} = 2\frac{t_l - t_p}{t_l + t_p},$$

where: V_p velocity of shear wave polarized along the loading axis, V_l velocity of shear wave polarized perpendicularly to the loading axis, t_l time of flight of the shear wave polarized along the loading axis, t_p time of flight of the shear wave polarized perpendicularly to the loading axis.

The value of acoustic birefringence depends on various factors influencing the velocity of ultrasonic waves. One can indicate a material texture (preferred grain orientations), concentration and orientation of voids (if any), impurities and dislocation in the bulk of the material, and applied residual stresses in the material.

The advantage of acoustic birefringence measurements is the fact that there is no need to know the exact thickness of the element under test. They do not depend on temperature and they deliver information averaged over the element thickness. Because of these features, this technique has found a wide application in ultrasonic residual stress evaluation in the rims of mono-block railway wheels (Schramm et al, 1996).

Ultrasonic measurements were performed in five spots on each specimen along gauge length. In this way, in specimens exhibiting necking, the maximal value of acoustic birefringence coefficient could be found. Measurements were taken with a 5 MHz shear wave piezoelectric transducer coupled to the specimen surface by means of the viscous epoxy couplant. Spots for measurements performed with ultrasonic techniques on the specimen are presented in Fig. 6.43. The same specimens as those used in magnetic investigations were considered.

It is shown, that also selected ultrasonic damage sensitive parameters can be correlated with the mechanical ones. In our case we have found the acoustic birefringence

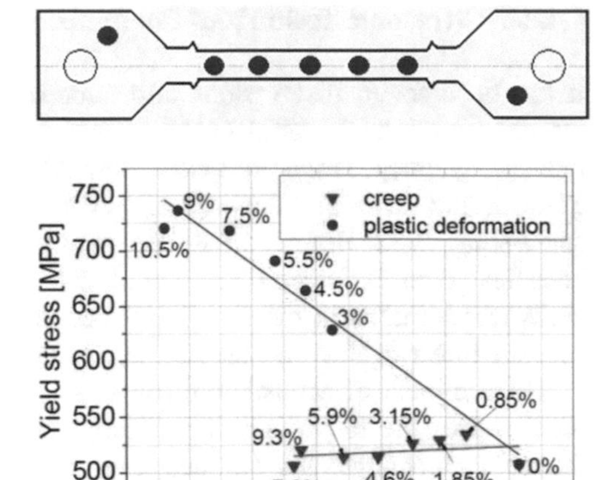

Fig. 6.43 Spots of ultrasonic measurements

Fig. 6.44 Variation of the yield stress for the P91 steel versus acoustic birefringence

coefficient as that which gives very promising results. Figure 6.44 well summarises some achievements in this area. It illustrates mutual relationship between the acoustic birefringence coefficient and yield stress. The numbers in figure denote the level of prior deformation. The yield stress of the P91 steel after deformation induced by creep almost does not change with the acoustic birefringence coefficient. In the case of prior plastic deformation the yield stress decreases linearly with an increase of the acoustic birefringence coefficient.

6.4.3 Application of Magnetic Techniques in Real Structural Elements for Rapid Inspection

The Barkhausen noise method, based on the measurement of voltage pulses generated by the magnetic domains moving due to the variable magnetic field (Buttle et al, 1986), is regarded as a promising research tool for assessing a degradation of the ferromagnetic materials used particularly in constructions for power engineering. Such a thesis is justified by the numerous papers describing the possibility of determining by this method a degree of material damage in various stages of creep (Palma et al, 2003; Mitra et al, 2007; Mohapatra et al, 2008; Makowska et al, 2014), fatigue (Palma et al, 2003; Sagar et al, 2005) or stress/strain assessment during plastic deformation due to tension (Stupakov et al, 2007; Piotrowski et al, 2009).

The 9Cr-1Mo (0.09% C) steel was tested under creep conditions ($p = 125$ MPa and $T = 600°$ C) (Mitra et al, 2007). The results showed, that the rms of the Barkhausen signal decreased in the primary creep, in the secondary period it dropped reaching the minimum, and then strongly increased. In the tertiary stage, its value increased slightly, taking almost linear relationship with respect to time.

The Barkhausen noise parameter was also determined during fatigue of the SAE8620 steel. It increased with the increasing number of fatigue cycles, and stress amplitude as well. Stupakov et al (2007) investigated the low-carbon steel CSN12013 (C = 0.03%) subjected to plastic deformation up to 23% approximately. It was found that with the increase of plastic deformation (up to 2.5%), the effective voltage of the Barkhausen emission increased, and subsequently decreased. Just mentioned examples of the research programs and their results show that the Barkhausen noise parameters can be successfully used to identify degradation progress of materials in power engineering elements. Applicability of this technique was also confirmed by our investigations in which a degree of degradation of the turbine blade material was tested. A relationship between the time of turbine blade degradation and Barkhausen noise signal was found (Makowska et al, 2017). The tests were carried out for the area of leading edge of the blade and those for the trailing edge part, Fig. 6.45. The representative results were summarized in Figs. 6.46-6.48. They show the results of the Barkhausen noise amplitude for blades after different values of the exploitation time. The values of this parameter indicate that for the blades after longest exploita-

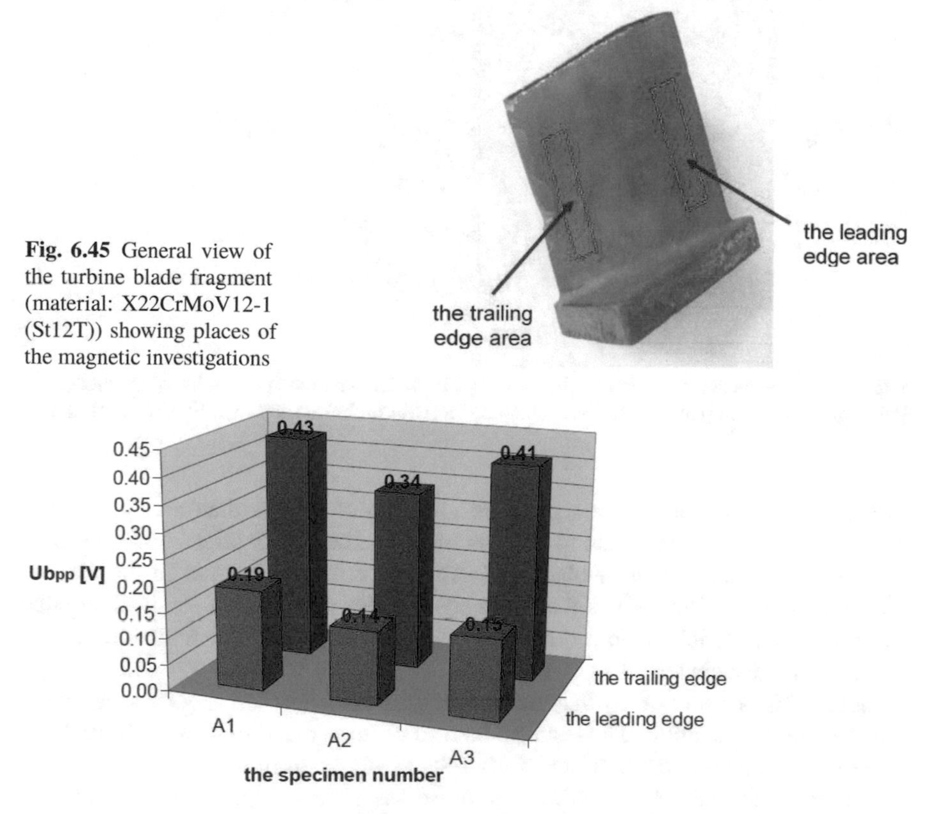

Fig. 6.45 General view of the turbine blade fragment (material: X22CrMoV12-1 (St12T)) showing places of the magnetic investigations

Fig. 6.46 The voltage difference (amplitude Ubpp) between maximum peak value of magnetic Barkhausen emission (Ub) and its background noise (Utb) for blades after 26400 [h] of exploitation

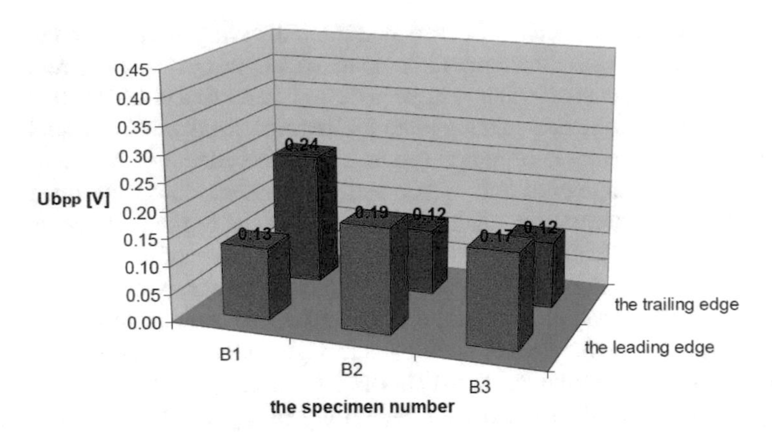

Fig. 6.47 The voltage difference (amplitude Ubpp) between maximum peak value of magnetic Barkhausen emission (Ub) and its background noise (Utb) for blades after 36100 [h] of exploitation

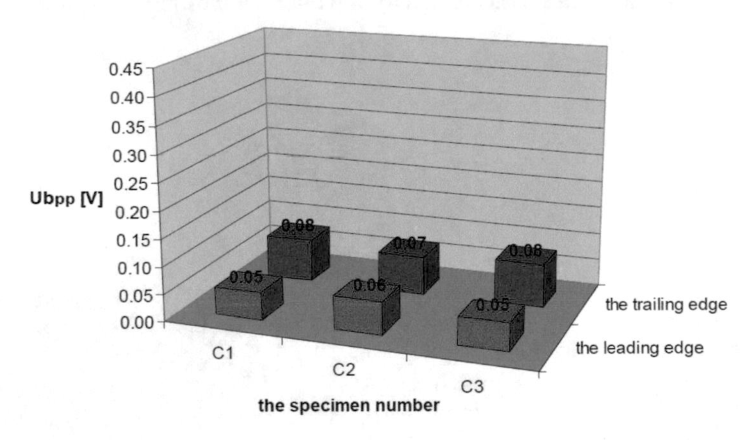

Fig. 6.48 The voltage difference (amplitude Ubpp) between maximum peak value of magnetic Barkhausen emission (Ub) and its background noise (Utb) for blades after 39800 [h] of exploitation

tion period considered here a significant lowering of the Barkhausen signal noise can be obseved in comparison to that obtained for the blades after 26400 hours of exploitation. Examination of blades A1-A3, Fig. 6.46, proves that in the case of blades without visible defects the Barkhausen noise captured on the leading edge was always proportionally lower than that on the trailing edge. This is also confirmed by the results for blades C1 - C3, Fig. 6.48.

In conclusion one can say that the Barkhausen noise method is quite sensitive to the material degradation. The tests enabled to observe the Barkhausen signal level decrease evaluated in the cracks vicinity located on the trailing edge of the blades. The lowest amplitude of the Barkhausen noise was observed for blades that suffered on structural degradation accumulated due to stress state variation around the defects in the form of blade material loss. The results achieved indicate that the Barkhausen noise method may in the future serve as an alternative technique for damage degree evaluation of selected components working in power engineering. At the current stage of the method development and taking into account a lack of the adequate standards of the presented concept, a further intensive investigations are needed for significantly larger number of cases.

6.5 Closing Remarks and Conclusions

The chapter presents selected results of damage development investigations due to various loading types obtained by means of destructive and nondestructive testing techniques.

The fatigue or creep tests for a range of different materials were interrupted for selected stages in order to assess a damage degree. As destructive methods the standard tensile tests were carried out after prestraining. Subsequently, an evolution of the selected tensile parameters was taken into account for damage identification.

The results of nondestructive tests show that the selected ultrasonic and magnetic parameters can be good indicators of material degradation and can help to locate the regions where material properties are changed due to prestraining.

Taking into account ultrasonic methods, in order to evaluate damage progress in specimens made of the steels, instead of velocity and attenuation measurement, the acoustic birefringence measurements were successfully applied. In the case of magnetic investigations for damage identification the measurements of the Barkhausen effect (HBE) and the magneto-acoustic emission (MAE) were applied. Both effects show that the magnetic properties are highly influenced by prior deformation, and moreover, they are sensitive not only to the magnitude of prior deformation, but also to the way it is introduced.

Among nondestructive methods the relatively novel optical techniques were used for damage initiation and further crack propagation monitoring. The results of damage monitoring during fatigue tests supported by Digital Image Correlation or Electronic Spackle Pattern Interferometry proved their great suitability for effective identification of places of damage initiation.

The results show that ultrasonic and magnetic damage sensitive parameters can be correlated with those coming from destructive tests. It is shown that good correlation of mechanical and selected non-destructive parameters identifying damage can be achieved for the materials in question.

This study suggests that experimental investigations concerning creep and fatigue problems should be based on the interdisciplinary tests, connecting parameters assessed by classical macroscopic destructive investigations, plus microscopic observations with parameters coming from the non-destructive investigations.

The chapter additionally presents simulation of fatigue crack initiation for cyclic loading within the nominal elastic regime. It is assumed that damage growth occurs due to action of mean stress and its fluctuations induced by crystalline grain inhomogeneity and free boundary effect. The simulations were supported by the results captured by ESPI system.

References

Andersson M (2013) The influence of notches on fatigue of heat treated sintered steel. In: International Powder Metallurgy Congress and Exhibition, Euro PM 2013 in Gothenburg on September 17, 2013

Augustyniak B (2003) Magnetomechanical effects and their application in nondestructive evaluation of materials (in Polish). Gdansk University of Technology, Gdańsk

Augustyniak B, Chmielewski M, Piotrowski L, Kowalewski Z (2008) Comparison of properties of magnetoacoustic emission and mechanical barkhausen effects for P91 steel after plastic flow and creep. IEEE Transactions on Magnetics 44(11):3273–3276, DOI 10.1109/TMAG.2008.2002525

Bailey RW (1935) The utilization of creep test data in engineering design. Proceedings of the Institution of Mechanical Engineers 131(1):131–349, DOI 10.1243/PIME_PROC_1935_131_012_02

Baldev R, Jayakumar T, Moorthy V, Vaidyanathan S (2001) Characterization of microstructures, deformation and fatigue damage in different steels using magnetic Barkhausen emission technique. Russian Journal of Nondestructive Testing 37(11):789–798, DOI 10.1023/A:1015847303916

Bennett JA, Weinberg JG (1954) Fatigue notch sensitivity of some aluminium alloys. Journal of Research of the National Bureau of Standards 52(5):235–245

Blaow A, Evans JT, Shaw BA (2007) The effect of microstructure and applied stress on magnetic Barkhausen emission in induction hardened steel. Journal of Materials Science 42(12):4364–4371, DOI 10.1007/s10853-006-0631-5

Boronski D (2007) Methods for examination of strain and stress under fatigue of material and structures (in Polish). Report, Institute for Sustainable Technologies - National Research Institute, Bydgoszcz-Radom

Brathe L (1978) Macroscopic measurements of creep damage in metals. Scandinavian Journal of Metallurgy 7(5):199–203

Buttle DJ, Briggs GAD, Jakubovics JP, Little EA, Scruby CB, Busse G, Sayers CM, Green RE, Ash EA, Scruby CB (1986) Magnetoacoustic and Barkhausen emission in ferromagnetic materials. Philosophical Transactions of the Royal Society of London Series A, Mathematical and Physical Sciences 320(1554):363–378, DOI 10.1098/rsta.1986.0124

Chen C (2004) General theory. ME631 15 pages, UAF

Chu TC, Ranson WF, Sutton MA, Peters WH (1985) Application of digital-image-correlation techniques to experimental mechanics. Experimental Mechanics 25(3):232–244, DOI 10.1007/BF02325092

DANTEC (2019) Dynamic info-materials. https://www.dantecdynamics.com/

Dietrich L, Kowalewski ZL (1997) Experimental investigation of an anisotropy in copper subjected to predeformation due to constant and monotonic loadings. International Journal of Plasticity 13(1):87 – 109, DOI 10.1016/S0749-6419(97)00002-8

Dietrich L, Grzywna P, Kukla D (2012) The use of optical methods to locate fatigue damage (in Polish). Przegląd Spawalnictwa 13:15–18

DuQuensay DL, Topper TH, Yu MT (1986) The effect of notch radius on the fatigue notch factor and the propagation of short cracks. In: Miller KJ, de los Rios ER (eds) The behaviour of short fatigue cracks, Mechanical Engineering Publications, London, pp 323–335

Durif E, Réthoré J, Combescure A, Fregonese M, Chaudet P (2012) Controlling stress intensity factors during a fatigue crack propagation using digital image correlation and a load shedding procedure. Experimental Mechanics 52(8):1021–1031, DOI 10.1007/s11340-011-9552-6

Fatemi A, Fang D, Zeng Z (2002) Notched fatigue behaviour under axial and torsion loads: experiment and predictions. In: 8th International Fatigue Congress, Stockholm, vol 3, pp 1905–1914

Fatemi A, Zeng Z, Plaseied A (2004) Fatigue behavior and life predictions of notched specimens made of qt and forged microalloyed steels. International Journal of Fatigue 26(6):663 – 672, DOI 10.1016/j.ijfatigue.2003.10.005

Fel D, Hsu DK, Warchol M (2001) Simultaneous velocity, thickness and profile imaging by ultrasonic scan. Journal of Nondestructive Evaluation 8(3):95–112, DOI 10.1023/A:1013550921673

Filippini M (2000) Stress gradient calculations at notches. International Journal of Fatigue 22(5):397 – 409, DOI 10.1016/S0142-1123(00)00010-4

Forster J, Theobald A, Engel S, Pasmann R (2012) Using optical measuring system for identification of material parameters for finite element analysis. In: 11. LS-DYNA, DYNAmore GmbH, Ulm, pp 1–9

Frost HJ, Ashby M (1982) Deformation-mechanism Maps. The Plasticity and Creep of Metals and Ceramics. Pergamon Press, Oxford, New York, Sydney

GOM (2007) ARAMIS user manual software

Gower MR, Shaw RM (2010) Towards a planar cruciform specimen for biaxial characterisation of polymer matrix composites. In: Advances in Experimental Mechanics VII, Trans Tech Publications Ltd, Applied Mechanics and Materials, vol 24, pp 115–120, DOI 10.4028/www.scientific.net/AMM.24-25.115

Gungor S (2009) Moire interferometry. In: Eaton-Evans J, Dulie-Barton JM, Burguete RL (eds) Modern Stress and Strain Analysis. A State of the Art Guide to Measurement Techniques, British Society for Strain Measurement, pp 8–9

Hayhurst D (1972) Creep rupture under multi-axial states of stress. Journal of the Mechanics and Physics of Solids 20(6):381 – 382, DOI 10.1016/0022-5096(72)90015-4

Jiles D (1998) Introduction to Magnetism and Magnetic Materials, 2nd edn. Taylor and Francis Group, New York

Kamaya M, Kawakubo M (2011) A procedure for determining the true stress–strain curve over a large range of strains using digital image correlation and finite element analysis. Mechanics of Materials 43(5):243 – 253, DOI 10.1016/j.mechmat.2011.02.007

Kim CS, Park IK, Jhang KY (2009) Nonlinear ultrasonic characterization of thermal degradation in ferritic 2.25Cr–1Mo steel. NDT & E International 42(3):204 – 209, DOI 10.1016/j.ndteint.2008.09.002, 2nd International Conference on Advanced Nondestructive Evaluation

Kopeć M, Grzywna P, Kukla D, Kowalewski Z (2012) Evaluation of the fatigue damage development using ESPI method. Inzynieria Materiałowa 4(212):1–4

Kowalewski ZL (2005) Creep of Metals – Experiment and Modeling (in Polish). IPPT PAN, Warsaw

Kwan MM, Ono K, Shibata M (1984) Magnetomechanical acoustic emission of ferromagnetic materials at low magnetization levels (type I behaviour). Journal of Acoustic Emisssion 3:144–156

Lanning DB, Haritos GK, Nicholas T (1999) Influence of stress state on high cycle fatigue of notched Ti-6Al-4V specimens. International Journal of Fatigue 21:S87 – S95, DOI 10.1016/S0142-1123(99)00059-6

Lord JD (2009) Digital image correlation (dic). In: Eaton-Evans J, Dulie-Barton JM, Burguete RL (eds) Modern Stress and Strain Analysis. A State of the Art Guide to Measurement Techniques, British Society for Strain Measurement, pp 14–15

Makowska K (2014) Methodology for assessing the state of damage to materials subjected to laboratory simulated operational loads (in Polish). PhD Thesis, Instytut Podstawowych Problemów Techniki PAN, Warsaw

Makowska K, Kowalewski ZL (2015) Possibilities of using Barkhausen noise to assess microstructure and mechanical properties of materials (in Polish). Energetyka - Problemy energetyki i gospodarki paliwowo-energetycznej 736(10):664–667

Makowska K, Kowalewski ZL, Augustyniak B, Piotrowski L (2014) Determination of mechanical properties of P91 steel by means of magnetic Barkhausen emission. Journal of Theoretical and Applied Mechanics 52(1):181–188

Makowska K, Kowalewski ZL, Ziółkowski P, Badur J (2017) Assessment of damage level of turbine blades using the Barkhausen noise signal (in Polish). Energetyka-Problemy energetyki i gospodarki paliwowo-energetycznej 760(10):638–641

Marténez-Ona R, Pérez MC (2000) Research on creep damage detection in reformer tubes by ultrasonic testing. In: Proc. 15 WCNDT Roma 2000

Maruno Y, Miyahara H, Noguchi H, Ogi K (2003) Notch size effects in the fatigue characteristics of Al-Si-Cu-Mg cast alloy. Journal of the Japan Institute of Metals 67(7):331–335, DOI 10.2320/jinstmet1952.67.7_331

Mazdumar PK, Lawrence Jr FV (1981) An analytical study of the fatigue notch size effect. A report of the fracture control program, College of Engineering, University of Illinois, Urbana, Illinois

Milke JG, Beuth JL, Biry NE (2000) Notch strengthening in titanium alumini dies under monotonic loading. Experimental Mechanics 40(4):415–424, DOI 10.1007/BF02326488

Mitra A, Mohapatra JN, Swaminathan J, Ghosh M, Panda AK, Ghosh RN (2007) Magnetic evaluation of creep in modified 9Cr–1Mo steel. Scripta Materialia 57(9):813 – 816, DOI 10.1016/j.scriptamat.2007.07.004

Mohapatra J, Ray A, Swaminathan J, Mitra A (2008) Creep behaviour study of virgin and service exposed 5Cr–0.5Mo steel using magnetic Barkhausen emissions technique. Journal of Magnetism and Magnetic Materials 320(18):2284 – 2290, DOI 10.1016/j.jmmm.2008.04.152

Mróz Z, Seweryn A, Tomczyk A (2005) Fatigue crack growth prediction accounting for the damage zone. Fatigue & Fracture of Engineering Materials & Structures 28(1-2):61–71, DOI 10.1111/j.1460-2695.2004.00829.x

Narayan R, Green Jr RE (1975) Ultrasonic attenuation monitoring of fatique damage in nuclear pressure vessel steel at high temperature. In: Materials Evaluation, 25-26 February 1975

Neuber H (1961) Theory of Notch Stresses. Office of Technical Services, U.S. Department of Commerce, Washington, D.C.

Neuber H (2001) Kerbspannungslehre, 4th edn. Klassiker der Technik, Springer, Berlin, Heidelberg

Ogi H, Minami Y, Aoki S, Hirao M (2000) Contactless monitoring of surface-wave attenuation and nonlinearity for evaluating remaining life of fatigued steel. In: Proc. 15 WCNDT Roma 2000

Ohtani T, Ogi H, Hirao M (2006) Evolution of microstructure and acoustic damping during creep of a Cr–Mo–V ferritic steel. Acta Materialia 54(10):2705 – 2713, DOI 10.1016/j.actamat.2006.02.010

Olszak W (1965) Theory of Plasticity (in Polish). PWN, Warsaw

O'Sullivan D, Cotterell M, Cassidy S, Tanner DA, Mészáros I (2004) Magneto-acoustic emission for the characterisation of ferritic stainless steel microstructural state. Journal of Magnetism and Magnetic Materials 271(2):381 – 389, DOI 10.1016/j.jmmm.2003.10.004

Palma ES, Júnior AA, Mansur TR, Pinto JMA (2003) Fatigue damage in AISI/SAE 8620 steel. In: Proceedings of COBEM 2003, 17th International Congress of Mechanical Engineering, 10-14 November, São Paulo, Brazil

Patorski K (2005) Laser Interferometry (in Polish), Oficyna Wydawnicza Politechniki Warszawskiej, Warsaw, pp 214–261

Peterson RE (1959) Analytical approach to stress concentration effect in fatigue of aircraft materials. In: Proceedings on Fatigue of Aircraft Structure, no. 59-507 in WADC Technical Report, pp 273–299

Pierron S (2009) Digital speckle pattern interferometry. In: Eaton-Evans J, Dulie-Barton JM, Burguete RL (eds) Modern Stress and Strain Analysis. A State of the Art Guide to Measurement Techniques, British Society for Strain Measurement, pp 14–15

Pilkey WD (1997) Peterson's Stress Concentration Factors. Wiley, New York

Piotrowski L, Augustyniak B, Chmielewski M, Tomáš I (2009) The influence of plastic deformation on the magnetoelastic properties of the CSN12021 grade steel. Journal of Magnetism and Magnetic Materials 321(15):2331 – 2335, DOI 10.1016/j.jmmm.2009.02.028

Pluvingae G (2001) Notch effects in fatigue and fracture. In: Pluvingae G, Gjonaj M (eds) Notch Effects in Fatigue and Fracture, Springer Science + Business

Media, Dordrecht, NATO Science Series (Series II: Mathematics, Physics and Chemistry), vol 11, pp 1–22

Qian G, Hong Y, Zhou C (2010) Investigation of high cycle and very-high-cycle fatigue behaviors for a structural steel with smooth and notched specimens. Engineering Failure Analysis 17(7):1517 – 1525, DOI 10.1016/j.engfailanal.2010.06.002

Sablik MJ, Augustyniak B (1999) Magnetic methods of nondestructive evaluation. In: Wiley Encyclopedia of Electrical and Electronics Engineering, Wiley, DOI 10.1002/047134608X.W4552

Sagar SP, Parida N, Das S, Dobmann G, Bhattacharya DK (2005) Magnetic Barkhausen emission to evaluate fatigue damage in a low carbon structural steel. International Journal of Fatigue 27(3):317 – 322, DOI 10.1016/j.ijfatigue.2004.06.015

Schneider E (1995) Ultrasonic birefringence effect — Its application for materials characterisations. Optics and Lasers in Engineering 22(4):305 – 323, DOI 10.1016/0143-8166(94)00032-6

Schramm RE, Szelązek J, Clark Jr AV (1996) Ultrasonic measurement of residual stress in the rims of inductively heated railroad wheels. Material Evaluation 54:929–934

Siebel E, Stieler M (1955) Ungleichformige Spannungsverteilung bei schwingender Beanspruchung. VDI Zeitschrift 97(5):121–126

da Silva BL, Ferreira JLA, Araújo JA (2012) Influence of notch geometry on the estimation of the stress intensity factor threshold by considering the Theory of Critical Distances. International Journal of Fatigue 42:258 – 270, DOI 10.1016/j.ijfatigue.2011.11.020, fatigue Damage of Structural Materials VIII

Stupakov O, Pal'a J, Tomáš I, Bydžovský J, Novák V (2007) Investigation of magnetic response to plastic deformation of low-carbon steel. Materials Science and Engineering: A 462(1):351 – 354, DOI 10.1016/j.msea.2006.02.475, international Symposium on Physics of Materials, 2005

Szelązek J, Mackiewicz S, Kowalewski ZL (2009) New samples with artificial voids for ultrasonic investigation of material damage due to creep. NDT & E International 42(2):150 – 156, DOI 10.1016/j.ndteint.2008.11.004

Szymczak T (2018) Investigations of material behaviour under monotonic tension using a digital image correlation system. Journal of Theoretical and Applied Mechanics 56(3):857–871, DOI 10.15632/jtam-pl.56.3.857

Szymczak T, Grzywna P, Kowalewski ZL (2013) Modern methods for determining the strength properties of construction materials (in Polish). Transport Samochodowy 1:79–104

Topper T, M Wetzel R, Dean Morrow J (1967) Neuber's rule applied to fatigue of notched specimens. Report NAEC-ASL-1114, Aeronautical Structures Laboratory

Toussaint F, Tabourot L, Vacher P (2008) Experimental study with a digital image correlation (DIC) method and numerical simulation of an anisotropic elastic-plastic commercially pure titanium. Archives of Civil and Mechanical Engineering 8(3):131 – 143, DOI 10.1016/S1644-9665(12)60168-X

Trąmpczyński W, Kowalewski ZL (1986) A tension-torsion testing technique. In: Science EA (ed) Proc. Symp. "Techniques for multiaxial creep testing", London and New York, pp 79–92

Ustrzycka A, Mróz Z, Kowalewski ZL (2017) Experimental analysis and modelling of fatigue crack initiation mechanisms. Journal of Theoretical and Applied Mechanics 55(4), DOI 10.15632/jtam-pl.55.4.1443

Vial-Edwards C, Lira I, Martinez A, Monzenmayer M (2001) Electronic speckle pattern interferometry analysis of tensile tests of semihard copper sheets. Experimental Mechanics 41(1):58–61, DOI 10.1007/BF02323105

Wahl AM, Beuwkes Jr R (1930) Stress concentration produced by holes and notches. Trans ASME APM-56-11:617–623

Westergaard HM (2014) Theory of Elasticity and Plasticity, Harvard Monographs in Applied Science, vol 3. Harvard University Press, Cambridge, MA

Whaley RE (1964) Fatigue and static strength of notched and un-notched aluminium alloy and steel specimens. Experimental Mechanics 2(11):329–334, DOI 10.1007/BF02326137

Zakharov VA, Ul'yanov AI, Gorkunov ES, Velichko VV (2013) Coercive-force hysteresis of carbon steels during elastic cyclic tensile deformation. Russian Journal of Nondestructive Testing 49(5):260–269, DOI 10.1134/S1061830913050070

Chapter 7
Heat Transfer Analysis in the Strapdown Inertial Unit of the Navigation System

Sergiy Yu. Pogorilov, Valeriy L. Khavin, Konstantin Naumenko, and Kyrill Yu. Schastlivets

Abstract This paper proposes an approach to modeling the temperature field of a strapdown inertial unit that is part of an inertial navigation system based on fiber-optic gyroscopes. A design scheme, a mathematical and finite element model for calculating the temperature field for the strapdown inertial unit has been developed. Results of numerical simulations including the effect of changes in external temperature on the temperature field in the device and temperature gradients at specified points of the device are presented.

Key words: Strapdown inertial unit · Navigation system · Heat transfer · Temperature gradients

7.1 Introduction

Currently, strapdown form inertial navigation systems based on fiber-optic gyroscopes (FOG) are widely used in the aerospace technology. The introduction of these systems requires the implementation of a wide range of studies aimed in identifying boundary opportunities, assessing accuracy and reliability, and other performance characteristics. In connection with the high sensitivity of FOG to the effect of temperature changes, the task of accurately determining the temperature field inside the devices, as well as the temperature change over time, is relevant and practically important.

Sergiy Yu. Pogorilov · Valeriy L. Khavin · Kyrill Yu. Schastlivets
Department of Continuum Mechanics and Strength of Materials, National Technical University "Kharkiv Polytechnic Institute", 61002 Kharkiv, Ukraine,
e-mail: ark95@ukr.net, vkhavin@kpi.kharkov.ua, sti66@mail.ru

Konstantin Naumenko
Institut für Mechanik, Otto-von-Guericke-Universität Magdeburg, 39106 Magdeburg, Germany,
e-mail: konstantin.naumenko@ovgu.de

H. Altenbach et al. (eds.), *Plasticity, Damage and Fracture in Advanced Materials*, Advanced Structured Materials 121,
https://doi.org/10.1007/978-3-030-34851-9_7

As part of the joint scientific research of National Technical University "Kharkiv Polytechnic Institute" and the company "HARTRON-ARKOS" to create a strapdown inertial navigation system (SINS) based on fiber-optic gyroscopes OIUS501, it became necessary to study thermal processes with a specific configuration of the SINS assembly. This paper is devoted to modeling the temperature field of a strapless inertial unit (SIN), which is part of the SINS to ensure the minimum temperature difference on the FOG platform in test and operating modes.

High performance requirements are set for modern fiber-optic gyroscopes, the most important of which is to ensure the FOG inertial accuracy of less than 0.01 degrees/hour under conditions of significantly varying temperatures in the range [-60°C – +60°C] (Dzhashitov and Pankratov, 2014). As a result, it is necessary to take into account the quantitative effect of temperature on the performance of FOG (Dzhashitov et al, 2014). Essentially important is the requirement to take into account thermal effects, leading to the emergence of the so-called thermal drift (fictitious, thermally induced changes in instrument readings) and thermal deformations of SINS parts (Dzhashitov and Pankratov, 2014; Dzhashitov et al, 2014)

To ensure the required accuracy of FOG, both passive methods of dealing with the influence of the temperature field (thermal compensation, special methods of winding the fiber of the FOG coil, thermal shunting, structural improvements of the FOG coil, etc.) (Shen and Chen, 2012; Shan, 2009; Zhang et al, 2012) and active methods consisting in the creation of multi-circuit reversible thermal control systems for both individual FOGs and the entire SINS (Dzhashitov and Pankratov, 2014; Zhang et al, 2015) are used in practice.

To improve the accuracy of measuring the angular velocity of fiber-optic and laser gyroscopes, an approach is used based on the post-processing of angular velocity measurements according to an algorithmic model called the temperature model of measurement errors. In this case, the approximation of the dependence of measurement errors (drifts) in the form of a third-degree polynomial has become widespread (Diesel and Dunn, 1996). In Breslavsky et al (2012), for the fiber-optic gyroscope OIUS501, it was proposed to determine the temperature gradient by calculating the difference between the readings of the temperature sensor at a given point and the temperature sensor located at another point. The same approach is used in the present work to numerically estimate the gradient of the temperature field of the original SIN system.

The aim of this study is to simulate the temperature field of the SIN device, and to determine the conditions that ensure minimum temperature differences on the FOG platform in the testing regime and during the operation of the system. To achieve this goal it is necessary to solve the following problems:

- Develop a design scheme and a finite element model of the SIN device
- Simulate the effect of changes in external temperatures on the temperature field of the SIN device.
- Determine temperature gradients at specified points of the device

7.2 Geometrical and Physical Process Modeling

The geometric model of the SIN is presented in Figs. 7.1 and 7.2. During the operation of SIN, the following heat sources can be considered (Fig. 7.3):

1 Three sources of the emulator of the electronics unit, which are three planes (each plane with a surface of heat generation of 85 cm^2). Every source generates heat with a power of 12 W;
2 Four sources of FOG (in each FOG there are two heat-generating surfaces), having the following heat generation characteristics:
   a the mounting surface (6.27 cm^2) of the emitting laser based on FOG with the power of 1.2 W

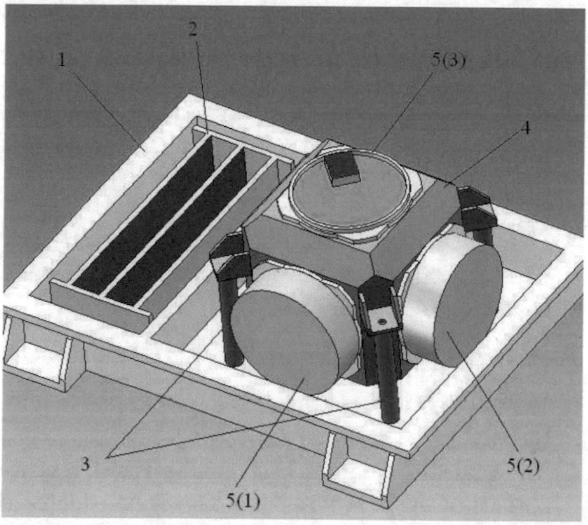

Fig. 7.1 Geometric model including retracted side panels, the upper FOG cover and the cover of the gyroscope number 3: 1 - heat sink base, 2 - emulator of the electronics unit, 3 - four insulating sleeves, 4 FOG platform, 5 - four FOG emulators, each of which consists of a FOG base and a FOG cover

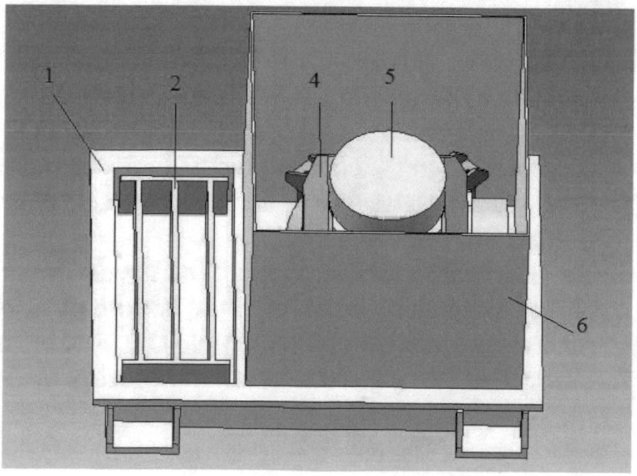

Fig. 7.2 Geometric model without top FOG compartment cover: 1 - heat sink base, 2 - emulator of the electronics unit, 4 FOG platform, 5 - four FOG emulators, each of which consists of a FOG base and a FOG cover, 6 - side heat receiving panels and the top cover of the FOG compartment (not shown in the figures)

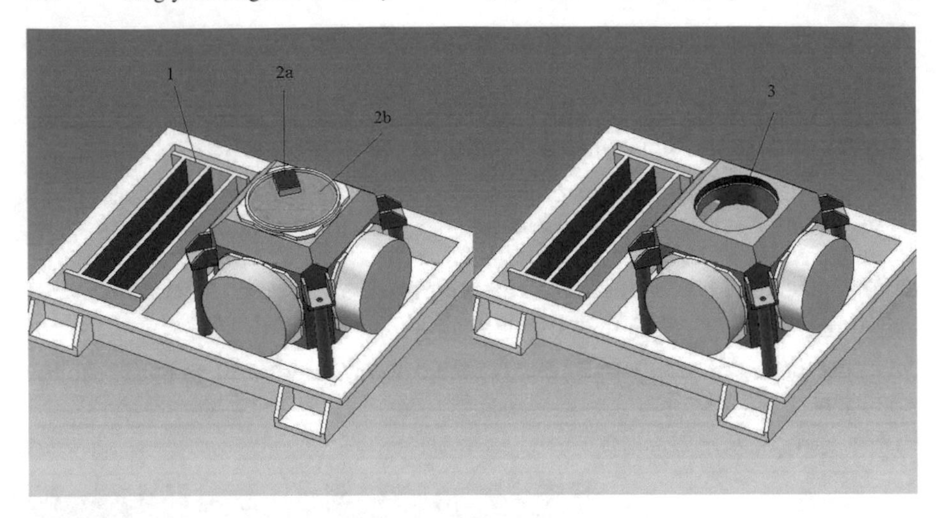

Fig. 7.3 Heat sources (the meaning of the numbers is given in the text)

b the rest of the inner surface of the base FOG (71.74 cm^2) with the power of 0.6 W

3 Four elements of the thermal stabilization system (heat-release surface area 27 cm^2) located on the FOG platform under the FOG installation sites. Each element has a power of 26 W

The physical heat transfer process was modeled as follows:

1. Heat flow from the cells of the electronics unit to the heat sink base is modeled by conductive heat transfer
2. The heat sink from the FOG platform to the heat-receiving panels and the inner surface of the heat-removing base is considered due to radiation from the FOG platform to the heat-receiving panels and the inner surface of the heat-removing base
3. Heat transfer between the heat-receiving panels and the heat-dissipating base is assumed by conduction at the fixing points of the panels to the base
4. Heat transfer between all components of the FOG platform is modeled by conduction through the contact points of the components
5. Conductive heat transfer at the contact points of system elements is simulated by an ideal thermal contact
6. The heat exchange of the device with the environment is modeled by the function of the temperature variation of the bottom plane of the heat sink base over time (interaction with the object on which the SIN is installed).
7. Convective heat transfer is ignored because of the intended use of FOG in vacuum

7.3 Mathematical Framework and Solution Method

The analysis of the heat transfer process and the temperature field is carried out by solving a transient heat conduction problem in a three-dimensional formulation. All elements of the FOG geometric model (Figs. 7.1 and 7.2) were considered as bulk bodies occupying a volume V with boundary Γ. The body is exposed to heat flux $q(x, y, z, t)$, where x, y, z are Cartesian coordinates and t is the time. The density of internal heat sources which can also be present inside the body is specified by $Q(x, y, z, t)$. The temperature distribution function inside the domain V is denoted by $T(x, y, z, t)$. To determine the unknown temperature distribution it is necessary to solve the following differential equation for the transient heat transfer (Nellis and Klein, 2009; Naumenko and Altenbach, 2016)

$$\frac{\partial}{\partial x}\left(k\frac{\partial T}{\partial x}\right)+\frac{\partial}{\partial y}\left(k\frac{\partial T}{\partial y}\right)+\frac{\partial}{\partial z}\left(k\frac{\partial T}{\partial z}\right)+Q-\rho c\frac{\partial T}{\partial t}=0, \tag{7.1}$$

where k is the coefficient of thermal conductivity, c is the specific heat capacity and ρ is the material density. For the uniqueness of the solution of the differential equation (7.1), the boundary and initial conditions should be supplemented. As initial conditions, the initial temperature distributions over the volume V are specified as follows

$$T(x, y, z, 0) = \overline{T}(x, y, z) \tag{7.2}$$

The following boundary conditions are considered:

1. Specified temperature values on the part of the surface of the body Γ_1 (condition of the 1st kind)

$$T(x, y, z, t)|_{\Gamma_1} = \overline{\overline{T}}(x, y, z, t) \tag{7.3}$$

   The function $\overline{\overline{T}}$ was set based on the experimental conditions. The surface Γ_1 in the model corresponds to the bottom surface of the heat sink base.
2. The specified values of the heat flux on the surface part of the body Γ_2 (condition of the 2nd kind)

$$-k\frac{\partial T(x, y, z, t)}{\partial n}\bigg|_{\Gamma_2} = \bar{q}_n, \tag{7.4}$$

   where n is the unit normal to the surface part Γ_2 and $\bar{q}_n$ is the given heat flux. The surface part Γ_2 corresponds to heat sources. The heat flux is determined by dividing the released thermal power to the areas of the respective surfaces.
3. Heat transfer by radiation. The following non-linear dependence of the heat flow rate on the temperature is applied

$$\dot{q} = \varepsilon_{eq} C_0 A_1 \left[\left(\frac{T_1}{100}\right)^4 - \left(\frac{T_2}{100}\right)^4\right], \quad \varepsilon_{eq} = \frac{1}{\frac{1}{\varepsilon_1} + \frac{A_1}{A_2}\left(\frac{1}{\varepsilon_2} - 1\right)}, \tag{7.5}$$

where ε_{eq} is the reduced emissivity coefficient, ε_1 the degree of blackness of the internal radiating surface, ε_2 the degree of blackness of the outer radiating surface, C_0 is the black body radiation coefficient, A_1 is the area of the internal radiating surface (elements of the FOG platform), A_2 is the area of the outer radiating surface (heat-receiving panels and the surface of the heat sink base bounded by them, as well as the compartment cover) T_1 is the temperature of the internal radiating surface, and T_2 is the temperature of the external radiating surface.

In the model, the radiation from the FOG platform to the heat-receiving panels as well as to the surface of the heat-dissipating base and the cover of the compartment limited by them is considered. To solve the heat transfer problem the finite element method and the ANSYS finite element code was applied. The finite element model of SIN consists of

- 679288 nodes with 679288 degrees of freedom (temperature)
- 362392 10-node tetrahedral elements
- 54533 auxiliary plane 4-node elements for the heat contact modeling

The material properties of the components of SIN are presented in Table 7.1. Heat radiation from external surfaces of the FOG platform and FOG covers with a blackness degree of 0.89 to the heat-receiving surfaces of the heat sink base, compartment covers and heat-receiving panels with a blackness degree of 0.90 is considered. Heat transfer through insulating sleeves was minimized by choosing the low thermal conductivity of the material. External thermal processes of the installation plane of the heat sink base are modeled by the function of temperature variation over time.

7.4 Results

The validation of the model by comparing results of simulation with experimental data is discussed in Pogorelov and Schastlivets (2005), where the temperature field of a similar device, namely a ring laser gyroscope (RLG), obtained by finite element analysis and a series of thermographic snapshots of working RLG was compared.

Table 7.1 Material properties of the components of SIN

Component	Material	Heat Capacity J/(kg K)	Thermal Conductivity W/(m K)	Density kg/m^3
Heat sink base and heat receiving panels	aluminium alloy	960	160	2700
Electronics emulator	aluminium alloy	960	160	2700
Thermal insulating sleeves	thermal insulation material	470	1	4540
FOG platform	aluminium alloy	960	160	2700
FOG cover and base	aluminium alloy	960	160	2700

The comparison parameters of the computational model and the experimental thermograms were the qualitative coincidence of the temperature field in the resonator, as well as the coincidence of the temperature values at the selected points.

Based on modeling the thermal mode of the device using a cyclogram of a constant external temperature, the analysis of the temperature field parameters in the "self-heating" mode ("self-heating time", overheating, temporary and alternative gradients) for a constant external temperature was performed. Furthermore possible locations of external temperature sensors for the purpose of simultaneous measurement of the temperature gradient over time were determined.

Heat sources can be in active and inactive states. For the computational experiment, the following heat sources were activated

- two sources of the emulator of the electronics unit, distant from the FOG platform compartment, and
- three sources of FOG: FOG1, FOG2, FOG3

During the numerical experiment, a non-stationary heat transfer problem with a duration of 14400 s was solved with a maximum time step of 300 s. The initial temperature of the SIN model is 298 K. The temperature of the heat sink base is constant and has the value 298 K. The temperature field at 14400 s is shown in Fig. 7.4. The results of numerical analysis can be summarized as follows

- Heat flow to the heat sink base by means of heat exchange by radiation at 14400 s is 5.022 W or 93% of the heat generation power (5.4 W)
- The magnitude of losses in the analysis of heat exchange by radiation is 0.277 W or 5% of the total heat generation power (5.4 W)
- Heat flow through heat insulating sleeves by means of heat transfer at 14400 s is equal to 0.0582 W or 1.08% of the total heat generation power

Thus, the total heat flux to the heat sink base is 94% of the total heat generation capacity. The error in the calculation of heat transfer by radiation, associated with the geometric imperfections of the model, is 5% of the total heat generation power.

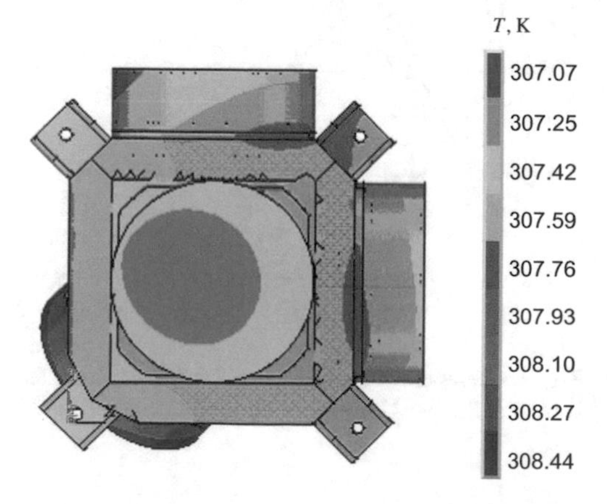

Fig. 7.4 Temperature field of the FOG platform and FOG sensors

Based on these data, it can be concluded that the SIN model is satisfactory with respect to the thermal balance parameters.

Figure 7.5 shows the change in temperature at the installation sites of temperature sensors as a function of time. The maximum temperature difference of the temperature field for the FOG platform and FOG sensors in steady state is approximately 1.5 K. The most warmed up places are the FOG mounting surfaces designated by FOG1-FOG3 (bottom), while the least are the upper surfaces of the FOG cover designated by FOG1-FOG3 (top). It should be noted that on the mounting feet for installing the FOG platform on the thermal insulating bushings, the temperature difference is approximately 0.5 degrees, which can lead to a slight rotation of the FOG platform.

Based on the obtained temperature values in the planned locations of installation of temperature sensors on the basis of active FOGs, the maximum values of overheating for FOG were obtained by subtracting the value of the constant external temperature from the temperature sensor reading, see Fig. 7.6. With the obtained temperature values in the expected locations of temperature sensors on the covers of all FOGs, an average value of overheating was obtained relative to the external temperature of the radiating surface of the FOG platform and FOG covers. The data are obtained by subtracting from the average temperature of 4 temperature sensors the value of the constant external temperature.

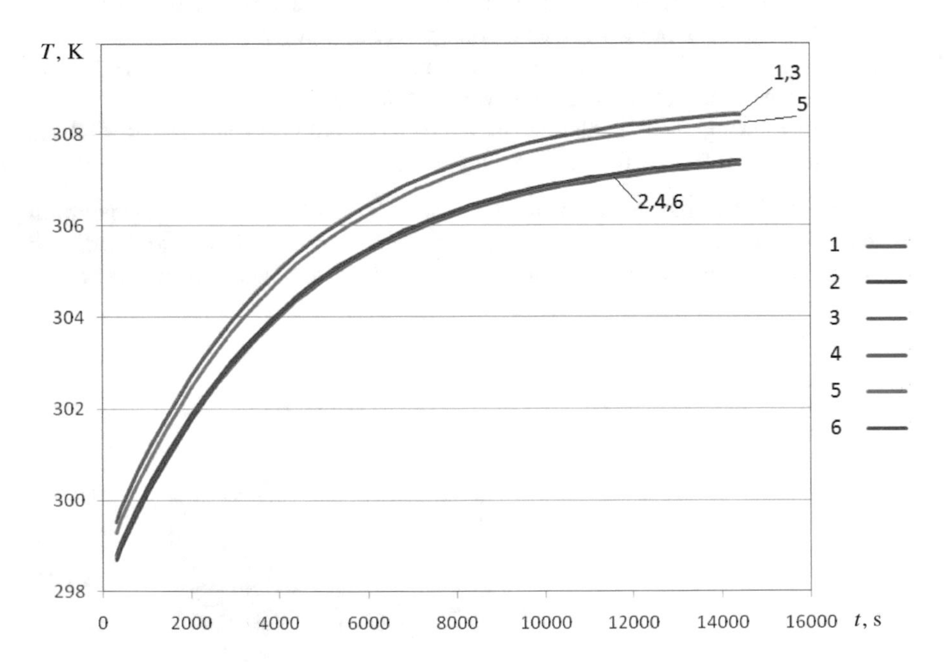

Fig. 7.5 Temperature variations in the installation sites of temperature sensors vs time. 1 – FOG1 (bottom), 2 – FOG1 (top), 3 – FOG2 (bottom), 4 – FOG2 (top), 5 – FOG3 (bottom), 6 – FOG3 (top)

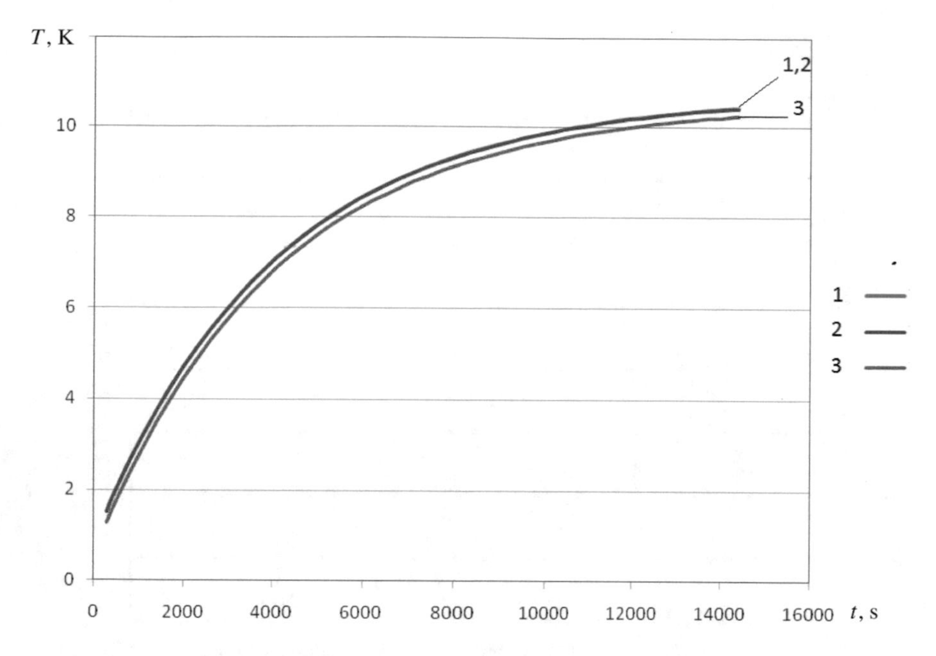

Fig. 7.6 Maximum overheating of FOG sensors. 1 – FOG1, 2 – FOG2, 3 – FOG3

At 14400 s, the average value of the overheating on the radiating surface is 9.3 K. According to the readings of temperature sensors on the basis of active FOG, temperature changes per minute, Fig. 7.7 were calculated to estimate the self-heating time. It can be seen that self-heating with a temperature rate of 0.02 K/min is reached in about 8000 s, and the level of 0.01 K/min is reached in about 11000 s, Fig. 7.7. At the time of 1800 s from switching on, temperature rate is 0.1 K/min. Based on the readings of the temperature sensors of active FOG, alternative temperature gradients were calculated by the method described in Dzhashitov et al (2014), as the difference between the sensor readings based on the active FOG and on the cover of the active FOG.

According to the results of the thermal mode of the device using a cyclogram of periodic changes in external temperature, the analysis of the temperature field parameters (self-heating time, overheating, temporary and alternative gradients) was performed. Furthermore possible locations of external temperature sensors to simultaneously measure the temperature gradient in time were determined. For the analysis, a non-stationary heat transfer problem with a duration of 29400 s was solved with a constant step of 120 s. The calculated experiment was conducted with initial conditions corresponding to the lower limit of the temperature range of operation. The initial temperature of the SIN model was 268 K. The temperature variation of the heat sink base was given according to the following law

$$T(t) = T_{\mathrm{A}} \sin\left(\frac{2\pi t}{t_{\max}} + \varphi\right) + T_0, \tag{7.6}$$

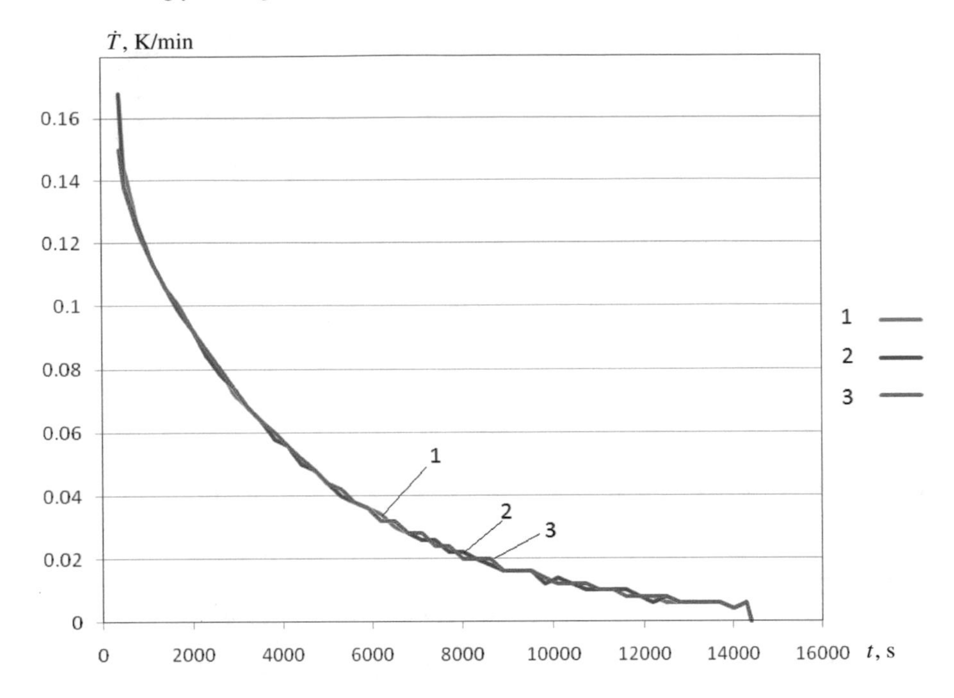

Fig. 7.7 Temperature rates in FOGs vs time. 1 – FOG1, 2 – FOG2, 3 – FOG3

where T_A, is the amplitude of temperature changes (20 K), t_{max} is the period of temperature changes, φ is the phase of temperature changes and T_0 is the mean temperature. In the analysis the following values were specified

$$T_A = 20\text{ K},\quad t_{max} = 5880\text{ s},\quad \varphi = \frac{3\pi}{4},\quad T_0 = 288\text{ K}$$

Figure 7.8 illustrates the temperature values as a functions of time presented on the bases of FOGs. The corresponding curves are numbered by 1 to 4. In addition, the external temperature in the node corresponding to the point on the lower surface of the heat sink base at the intersection of the diagonals between the attachment points of the insulating sleeves is plotted as a function of time (curve 5). Based on these results, it can be seen that at steady-state thermal regime, which occurs approximately after the third cycle of change in external temperature, the variation in the maximum temperature of the FOG does not exceed ±5 K relative to 300 K (27 °C). Therefore we may conclude that it is possible not to use the thermal stabilization system in the established thermal mode under considered service conditions.

Figure 7.9 shows temperature rates of the bases of FOGs as functions of time. In addition the rate of external temperature is plotted. It can be seen that the temperature rates of FOGs do not exceed 0.27 K per minute under steady thermal conditions and given operating conditions. This allows us to conclude the use of the thermal stabilization system in the established mode under the given operating conditions

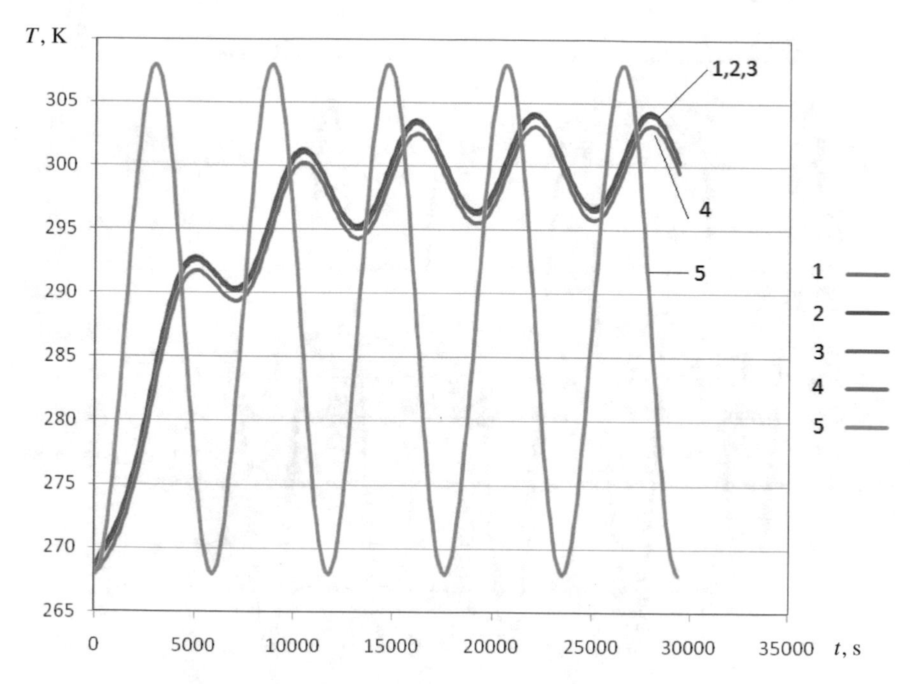

Fig. 7.8 Temperature variations vs time. 1 – FOG1 (bottom), 2 – FOG2 (bottom), 3 – FOG3 (bottom), 4 – FOG4 (bottom), 5 – external temperature

is not necessary. Figure 7.10 shows the temperature rates computed by numerical time differentiation for the base of FOG1 (curve 1) and for the external temperature (curve 3). In addition, the alternative approach differences proposed in Breslavsky et al (2012) is applied to compute temperature differences. Curve 2 is the result obtained from the temperature difference in the nodes on the basis and the cover of VOG1, while curve 3 is the result according to the difference of the temperature in the nodes on the basis and the outside temperature.

The numerical analysis was repeated with the initial conditions and the law of change in the external temperature corresponding to the upper limit of the temperature range of operation. The initial temperature of the SIN model was 308K. The results are illustrated in Figs. 7.11 and 7.12. From the obtained results, it can be concluded that the parameters of the temperature field in the steady state mode do not depend on the initial conditions. The time of occurrence of the established thermal regime is significantly less than in the case of initial conditions corresponding to the lower limit of the operating range.

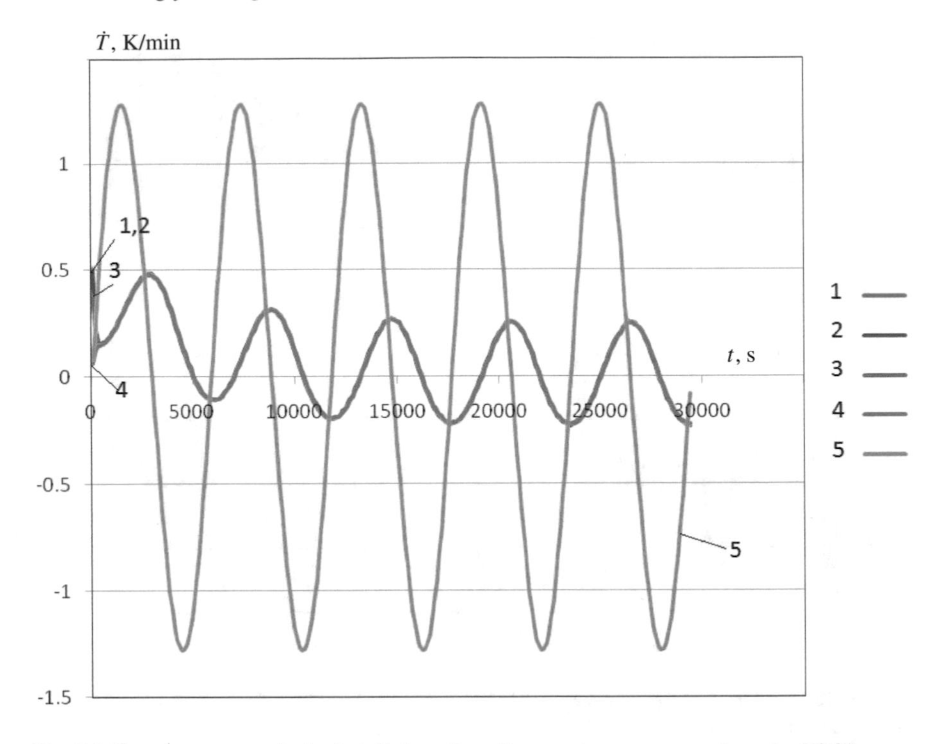

Fig. 7.9 Temperature rates in the installation sites of temperature sensors vs time. 1 – FOG1 (bottom), 2 – FOG2 (bottom), 3 – FOG3 (bottom), 4 – FOG4 (bottom), 5 – external temperature rate

7.5 Conclusions

This paper proposes an approach to modeling the temperature field of a strapdown inertial unit that is part of a strapdown inertial navigation system based on fiber-optic gyroscope. The geometrical, mathematical and finite element models for calculating the temperature field for a free-form inertial unit has been developed. Numerical simulations of the effect of changes in external temperature on the temperature field in the device and temperature rates at specified points of the device is given. It is revealed that the considered design of the SIN device with a heat sink from the FOG platform by means of radiation allows to obtain temperature field with a temperature difference of 2 K. This allows us to ensure a minimum level of temperature distortions of the FOG platform and reduce fluctuations of the external temperature at a rate of more than 8 K/ min to the level of 0.35 K/ min. In this case, FOG overheats in relation to the ambient temperature are up to 13 K. This should be taken into account when operating the instrument at an ambient temperature close to the upper operating limit.

The considered structural scheme of SIN separates the heat fluxes from the cells of the electronics and the FOG platform while ensuring the temperature of the base independent of the thermal conditions of the device. For all considered models

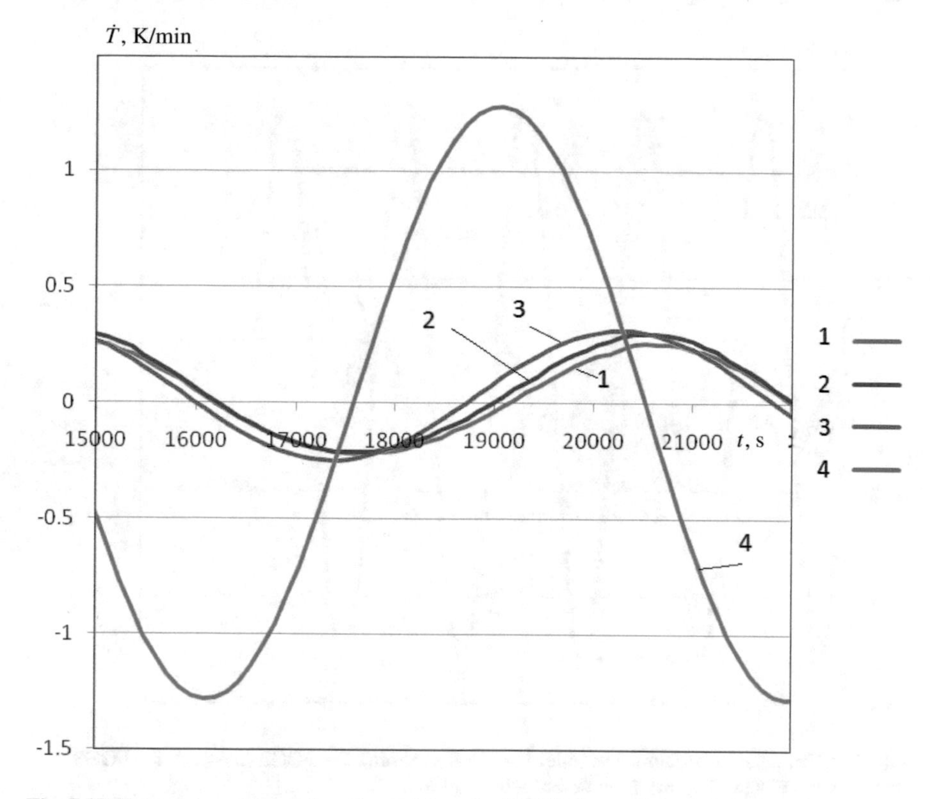

Fig. 7.10 Temperature rates as functions of time. 1 – FOG1 (bottom), 2 – FOG1 (bottom), 3 – FOG1 (bottom), 4 – external temperature rate

of external thermal conditions, the FOG operating temperatures and temperature gradients are in the zone of permissible operating characteristics.

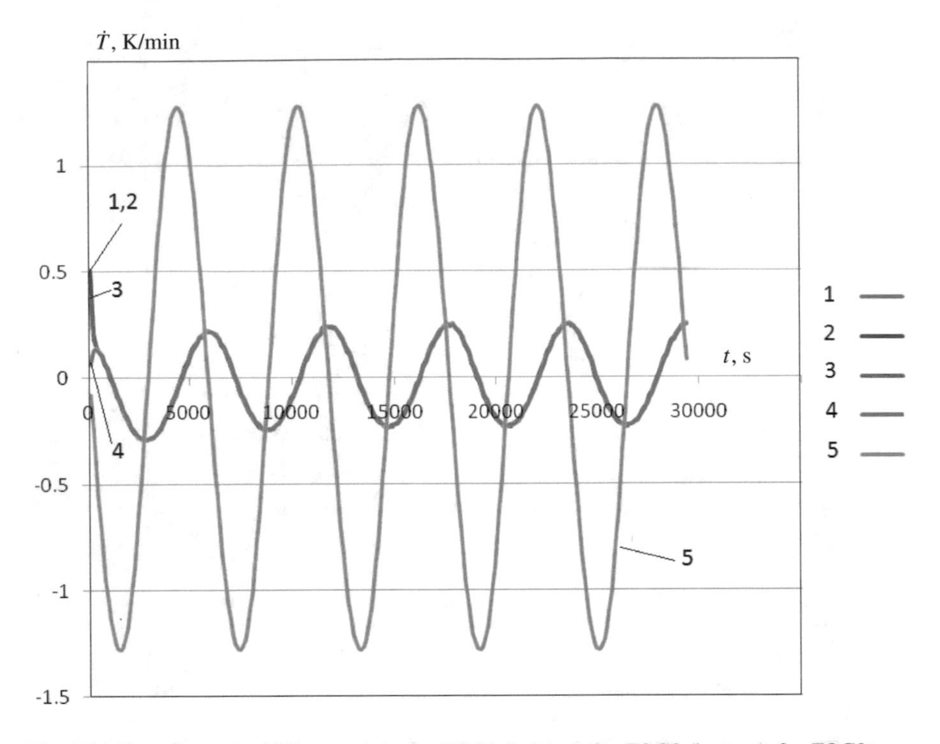

Fig. 7.11 Temperature variations vs time. 1 – FOG1 (bottom), 2 – FOG2 (bottom), 3 – FOG3 (bottom), 4 – FOG4 (bottom), 5 – external temperature

References

Breslavsky D, Pogorelov S, Schastlivets K, Batyrev B, Kuznetsov Y, Oleynik S (2012) Development of method to determine temperatur gradients in fiber-optic gyroscope OIUS501. Mechanika ta Mashinobuduvannya (1):90–101, (in Russian)

Diesel JW, Dunn GP (1996) Method for in-field updating of the gyro thermal calibration of an intertial navigation system. US Patent 5,527,003

Dzhashitov V, Pankratov V (2014) Control of temperature fields of a strapdown inertial navigation system based on fiber optic gyroscopes. Journal of Computer and Systems Sciences International 53(4):565–575

Dzhashitov V, Pankratov V, Golikov A (2014) Mathematical simulation of temperature field control of the strapdown inertial navigation system based on optical fiber sensors. Journal of Machinery Manufacture and Reliability 43(1):75–81

Naumenko K, Altenbach H (2016) Modeling High Temperature Materials Behavior for Structural Analysis: Part I: Continuum Mechanics Foundations and Constitutive Models, Advanced Structured Materials, vol 28. Springer

Nellis G, Klein S (2009) Heat Transfer. Cambridge University Press

Pogorelov S, Schastlivets K (2005) Refinement of the computational model of a ring laser gyroscope based on experimental data. Vestnik NTU KhPI (47):153–158,

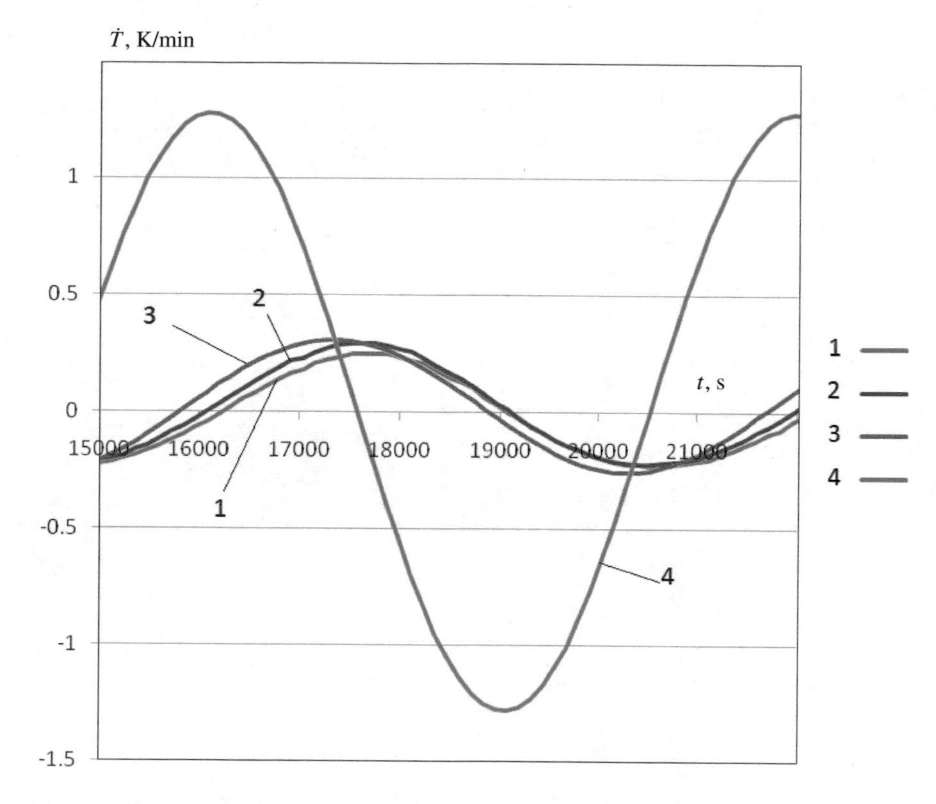

Fig. 7.12 Temperature rates as functions of time. 1 – FOG1 (bottom), 2 – FOG1 (bottom), 3 – FOG1 (bottom), 4 – external temperature rate

(in Russian)

Shan WXM (2009) Nonlinearity of temperature and scale factor modeling and compensating of FOG. Journal of Beijing University of Aeronautics and Astronautics 25:28–31

Shen C, Chen X (2012) Analysis and modeling for fiber-optic gyroscope scale factor based on environment temperature. Applied Optics 51(14):2541–2547

Zhang Y, Wang Y, Yang T, Yin R, Fang J (2012) Dynamic angular velocity modeling and error compensation of one-fiber fiber optic gyroscope (OFFOG) in the whole temperature range. Measurement Science and Technology 23(2):025,101

Zhang Y, Guo Y, Li C, Wang Y, Wang Z (2015) A new open-loop fiber optic gyro error compensation method based on angular velocity error modeling. Sensors 15(3):4899–4912

Chapter 8
Influence of Citric Acid Concentration and Etching Time on Enamel Surface Roughness of Prepared Human Tooth: *in vitro* Study

Evgeniy V. Sadyrin, Evgeniy A. Kislyakov, Roman V. Karotkiyan, Diana V. Yogina, Ekaterina G. Drogan, Michael V. Swain, Stanislav Yu. Maksyukov, Andrey L. Nikolaev, and Sergei M. Aizikovich

Abstract In the present paper the average surface roughness *Ra* of three human enamel specimens subjected to etching using citric acid of different concentrations was evaluated using atomic force microscopy (AFM). The specimens were obtained from a single tooth. The cut sections of enamel were mounted in epoxy resin, ground and polished. Each specimen was etched using one of the concentrations (0.5 wt. %, 3 wt. %, 5 wt. %) of citric acid. For each concentration five etching durations (1 s, 15 s, 30 s, 45 s, 60 s) were applied. The AFM images of similar size and resolution were evaluated for each sample after each etching duration. For each of the images *Ra* in three directions – horizontal, vertical and diagonal – was calculated and surface profiles constructed. The average value of *Ra* in three directions was taken as the absolute value of the roughness for a given surface. Dependence of the absolute value of the surface roughness upon the etching time for all samples was calculated.

8.1 Introduction

Teeth are exposed to a variety of acidic conditions during daily functioning. A very common source of exposure is citric acid associated with fruits or juice ingestion. Despite the exceptional strength properties of the human enamel covering the under-

Evgeniy V. Sadyrin, Evgeniy A. Kislyakov, Roman V. Karotkiyan, Ekaterina G. Drogan, Michael V. Swain, Stanislav Yu. Maksyukov, Andrey L. Nikolaev, Sergei M. Aizikovich
Don State Technical University, 1 Gagarin Sq., Rostov-on-Don, 344000, Russia,
e-mail: ghostwoode@gmail.com, evgenka_95@mail.ru, valker94@gmail.com, ekaterina.drogan@gmail.com, michael.swain@sydney.edu.au, andreynicolaev@eurosites.ru, saizikovich@gmail.com

Diana V. Yogina, Stanislav Yu. Maksyukov
Rostov State Medical University, 29 Nakhichevansky Lane, Rostov-on-Don, 344022, Russia,
e-mail: dianaturbina@mail.ru, kafstom2.rostgmu@yandex.ru

H. Altenbach et al. (eds.), *Plasticity, Damage and Fracture in Advanced Materials*, Advanced Structured Materials 121,
https://doi.org/10.1007/978-3-030-34851-9_8

lying dentine in the crown area if the tooth, acidic food and beverages are capable of its dissolution (Barbour et al, 2003; Zheng et al, 2009). It can result in substantial loss of the tooth surface. On the other hand, in clinical practice acid localized etching is widely used to achieve increasing bonding of restorative materials to the tooth surface. Etching enables partial dissolution and demineralization the inorganic matrix from the surface of the enamel, forming microporosities and microgrooves (Lopes et al, 2006; Moura et al, 2006; Retief and Denys, 1979; Torres-Gallegos et al, 2012; Zanet et al, 2006). This process results in increasing roughness of the enamel surface. Prior enamel etching is able to improve bond strength values of resin based bonding systems to enamel, compared to using of the same systems without prior etching (Erhardt et al, 2004; Landuyt et al, 2006). For these purposes a variety of different acids can be applied (Cardenas et al, 2018; van der Vyver et al, 1997).

Orthodontists also use acid to form retentive surfaces, which on the one hand improves the bracket fixation but on the other hand weakens the structure of the adjacent enamel prisms, and increases the permeability and biofilm adhesion of enamel and reduces its protective properties (Gordeeva et al, 2004). For scientific purposes judicious application of acid etching allows visualization of the structural elements of tooth enamel for further study.

Atomic-force microscopy (AFM) has been widely used to study the effects of acid etching on the enamel surface state. Sorozini et al (2018) studied the possibility of different forms of enamel preparation to influence the advent of artifacts during AFM observations. Watari (2005) made a research of the evolution of the tooth surface microgeometry applying three acid agents. Dong et al (2010) used AFM to study the capabilities of tricalcium silicate toothpaste to remineralize the acid etched human enamel in vitro. Loyola-Rodriguez et al (2009) compared the enamel surface roughness and absolute depth profile before and after using four different phosphoric acids using AFM. In the present research we used AFM to evaluate the average surface roughness Ra of four human enamel specimens from a single tooth. The specimens were subjected to etching using citric acid of different concentrations and for different durations. For each "acid concentration – etching duration" pair the values of Ra were quantitatively determined in three directions – horizontal, vertical and diagonal, surface profiles were built and AFM images of the surface were collected.

8.2 Materials and Methods

8.2.1 Sample Preparation

Three human enamel specimens were obtained from a maxillary molar, which was removed from a patient (male, 21 years old). The tooth was extracted in the dental department of Rostov State Medical University clinic for orthodontic reasons (ethics commitee of Rostov State Medical University approved the study, the patient provided informed consent). After removal, the tooth was kept in 1% NaClO solution for 10 min. Then the tooth was stored in Hanks Balanced Salt Solution (HBSS) at 4°C with thymol granules added to prevent fungal growth. The ratio of thymol to HBSS was 1:1000. The tooth was mounted in epoxy resin for the convenience of cutting. Cutting parameters on the Isomet 4000 machine (Buehler, Germany) were as follows: disk rotation speed 2500 rpm, sample feed rate 10 mm/min. Saw cooling was carried out by continuous supply of coolant cutting fluid (Cool 2, Buehler, USA). Two cuts were made in such a way that the second cut formed a slice of the middle part of the tooth containing a white spot lesion (the sample was moved relative to the disk using the micrometer built into the machine). Thus, after cutting two longitudinal sections containing sound enamel were obtained.

On the second step each of these two sections were additionally cut in a transversal direction perpendicular to the plane of the first cut. The cutting was carried out on the same machine and same parameters, however this time we used cubic boron nitride disc with the same coolant supply. As a result, three subsections were chosen to be the samples for the study. Each sample was again mounted in epoxy resin to enable subsequent grinding and polishing, which was carried out simultaneously for all three samples on a MetaServ 250 machine (Buehler, PRC). The loading on the sample at each of the stages of grinding and polishing was 10 N, the disk rotation speed was 100 rpm. Running water was used as a lubricant. The details on grinding parameters are shown in the Table 8.1. For polishing the same load and disk rotation speed were applied. The lubricant consisting of glycerin and propylene glycol mixed in 1:1 proportion was prepared. The details on polishing parameters are shown in the Table 8.2. Between each grinding and polishing step ultrasonic cleaning of each sample was made in distilled water (Sonorex RK 31, Bandelin, Germany) bath for 5 min (and for 10 min after the final polishing).

Table 8.1 Grinding parameters

Grinding step	Grit size	Abrasive paper	Grinding time
1	P800	Siawat 1913 SiC-based (Sia Abrasives, Switzerland)	60 s
2	P1500	Smirdex SiC-based (Smirdex, Greece)	30 s
3	P2500	Smirdex SiC-based (Smirdex, Greece)	30 s

Table 8.2 Polishing parameters

Polishing step	Particle size	Abrasive suspension	Polishing time	Lubricant	Cloth
1	6 μm	MetaDi	120 s	glycerin / propylene glycol	Trident
2	3 μm	MetaDi	300 s	glycerin / propylene glycol	Trident
3	1 μm	MetaDi	360 s	glycerin / propylene glycol	Veltex
4	0.05 μm	MasterPrep	480 s	distilled water	Veltex

8.2.2 Optical Microscopy

After sample preparation the tooth surfaces were examined using a Stemi 305 optical Greenough stereomicroscope (Zeiss, PRC) equipped with Axiocam 105 color video camera (Zeiss, Germany) in reflected light (Fig. 8.1). The enamel, dentine and dentine-enamel junction were clearly visible in the images indicating the sound state of the samples under research. Sample 1 was etched with acid of concentration 0.5 wt. % (pH = 2.3), sample 2 with concentration 3 wt. % (pH = 1.86), sample 3 with concentration 5 wt. % (pH = 1.71). The measurements of pH were conducted (Cobra4 Mobile-Link, PHYWE, Germany). Before pH measurements the sensor was calibrated in solutions of known pH. On each sample the etching was sequentially conducted for five durations: 1 s, 15 s, 30 s, 45 s, 60 s. For the 1 s etching a wet lint-free cloth with the acid briefly touched the surface of the sample. For the other four durations etching was carried out by dipping the sample in an appropriate acid solution. Following acid exposure ultrasonic cleaning in distilled water was made.

8.2.3 Atomic Force Microscpoy

The research of the enamel surface topography for each sample before etching and after each etching duration was conducted on AFM Nanoeducator (NT-MDT, Russia) equipped with a tungsten probe. The device worked in non-contact mode. For all the images obtained the resolution was 256x256 points and the scanning velocity was 20 μm/s. Positioning was performed using the optical USB microscope fixed on the tripod.

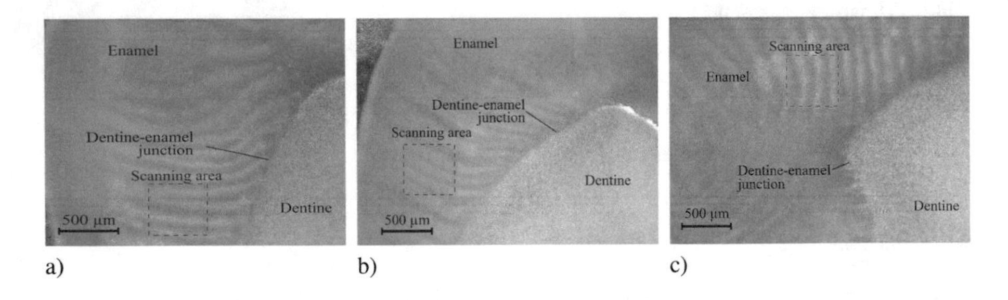

Fig. 8.1 Surfaces of the tooth areas with the selected area for AFM scanning of the prepared samples: a) 1, b) 2, c) 3

8.3 Results

8.3.1 Optical Observations

The enamel, dentine and dentine-enamel junction were clearly visible in the images that indicate the sound state of the samples under research (Fig. 8.1). Hunter-Schreger bands were also visualized on all the samples.

8.3.2 Atomic Force Microscpoy Surface Monitoring

Figure 8.2 demonstrates the surface of each of the three samples before etching and after 1 s etching. In Fig. 8.3 the observations of the surface of each of the samples after 15 s and 30 s etching is shown. Figure 8.4 shows the results after 45 s and 60 s etching, while in Fig. 8.5 the surface profiles for each of the samples on each etching step are demonstrated. The single crystallites of hydroxyapatite can be seen on the image obtained with the AFM Nano Compact (Phywe, Germany) and horizontal surface profile (Fig. 8.6) for the sample 1 after 60 s etching time (Fig. 8.7). Scanning field was 1.01 x 1.01 μm. The microscope was equipped with monocrystal Si probe with Al coating with the resonance frequency 190 ± 60 kHz and constant stiffness 48 N/m. The surface scanning was conducted in the dynamic mode with the scanning velocity 0.3 ms per line. The resolution along the axis was 1.1 nm. The quadratic mean noise level in the mode across the height axis was 0.5 nm.

8.3.3 Surface Roughness Measurement

Surface roughness *Ra* was measured on each etching step for each sample using Gwyddion software (Czech Metrology Institute, Czech Republic). The latter was determined from the AFM generated surface profile results shown in Figures 8.2–8.5. Due to the possible irregularity of the visualized structure we considered measurement of the roughness of based on a single direction single held roughness profile as not being completely correct. In the present paper, surface roughness was measured in three directions: horizontal, vertical, diagonal. In each of the directions five profiles were constructed. The average value of these five profiles was taken as absolute in this direction (Ra_{avg}). The average value of roughness in three directions is the absolute value of the roughness of a given surface (Ra_{absol}). The measurement results for samples 1-3 are presented in Tables 8.3–8.5 respectively (Tables 8.4 and 8.5 are shown in the shortcut form).

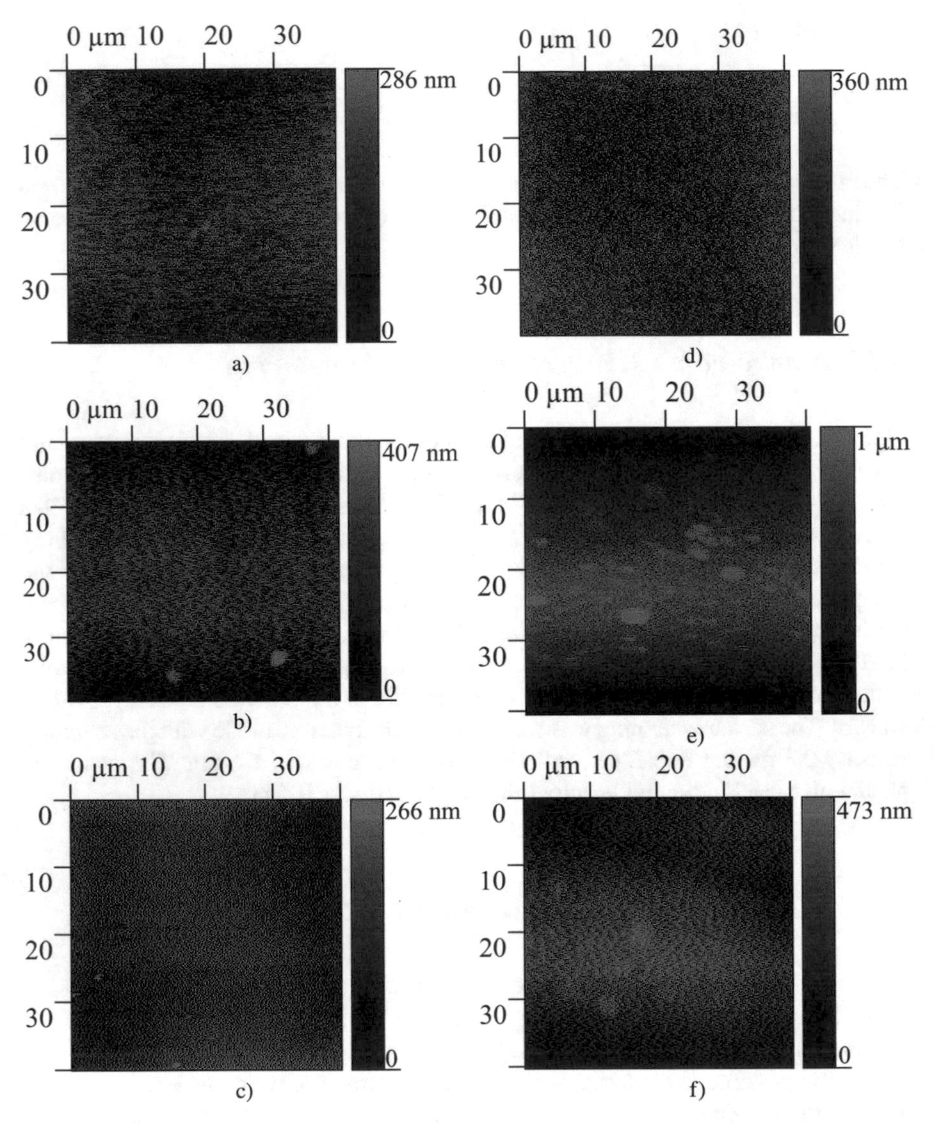

Fig. 8.2 AFM images of the surfaces before etching for the samples: a) 1, b) 2, c) 3, and after 1 second etching: d) 1, e) 2, f) 3. The images do not display details of microstructure

Table 8.3 Surface roughness Ra for the sample 1

t	n	Horizontal $Ra(n)$	Ra_{avg}	Vertical $Ra(n)$	Ra_{avg}	Diagonal $Ra(n)$	Ra_{avg}	Ra_{absol}
0	1	26.9	22.5	26.6	27.1	18.4	22.8	24.1
	2	26.3		27.1		21.3		
	3	20.8		26.6		21.3		
	4	21.0		28.7		28.4		
	5	17.3		26.5		24.3		
1	1	30.4	31.0	33.1	29.6	20.0	24.4	28.3
	2	31.5		28.7		27.0		
	3	30.0		25.8		20.7		
	4	27.8		28.9		30.4		
	5	35.3		31.3		23.8		
15	1	27.6	30.5	19.4	20.2	24.8	26.4	25.7
	2	30.1		17.0		32.8		
	3	31.6		20.4		21.8		
	4	35.6		18.4		32.1		
	5	27.6		26.0		20.5		
30	1	91.7	99.5	82.1	101.8	82.5	103.0	101.4
	2	153.0		125.3		103.6		
	3	107.4		110.9		147.8		
	4	67.0		110.1		110.2		
	5	78.5		80.4		70.9		
45	1	125.6	120.0	153.1	137.7	128.4	115.0	124.2
	2	137.9		133.8		115.9		
	3	133.2		86.4		126.4		
	4	104.3		153.3		126.6		
	5	99.2		161.8		77.6		
60	1	194.2	130.8	111.4	126.6	160.6	161.8	139.7
	2	142.0		145.6		150.6		
	3	94.5		123.8		130.3		
	4	133.1		134.4		187.0		
	5	90.1		117.9		180.6		

Table 8.4 Surface roughness Ra for the sample 2 (summary)

t	Horizontal Ra_{avg}	Vertical Ra_{avg}	Diagonal Ra_{avg}	Ra_{absol}
0	17.1	32.6	28.0	25.9
1	20.4	21.2	17.4	9.7
15	28.7	40.6	27.6	32.3
30	52.4	70.9	84.3	69.2
45	64.0	69.6	64.2	65.9
60	79.0	83.8	87.7	83.5

8.4 Discussion

Tracing the changes that developed on the surfaces of the samples during etching allowed us to make some specific remarks. The surface prior to etching (Fig. 8.2a, b, c) and after 1 s etching (Fig. 8.2d, e, f) demonstrated the presence of an almost

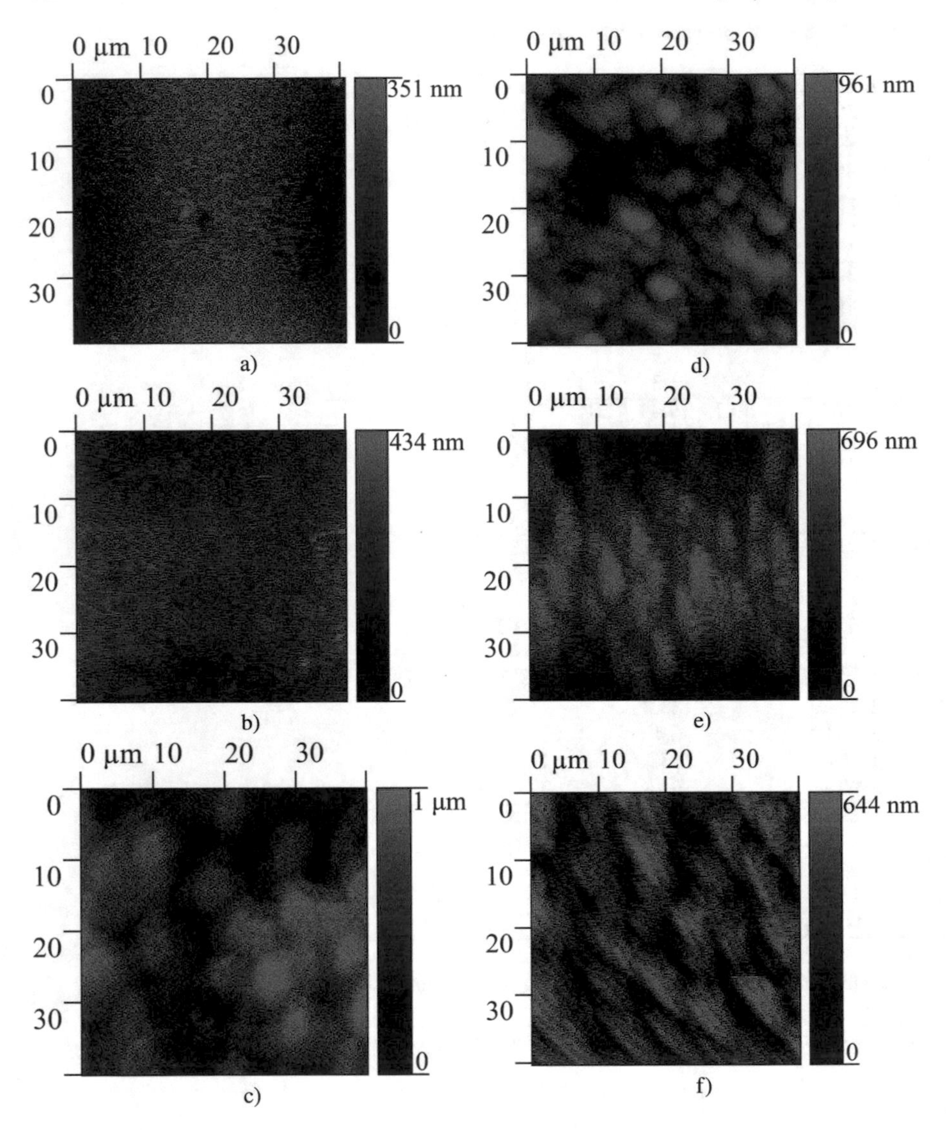

Fig. 8.3 AFM images of the surfaces after 15 seconds etching of the samples: a) 1, b) 2, c) 3 and after 30 seconds etching: d) 1, e) 2, f) 3. Qualitative change of the surface microgeometry

Table 8.5 Surface roughness Ra for the sample 3 (summary)

t	Horizontal Ra_{avg}	Vertical Ra_{avg}	Diagonal Ra_{avg}	Ra_{absol}
0	25.7	19.0	15.8	20.2
1	26.6	43.6	38.2	36.1
15	126.3	121.7	106.7	118.2
30	89.0	77.5	87.2	84.6
45	171.1	151.9	142.0	155.0
60	255.8	225.0	242.0	241.0

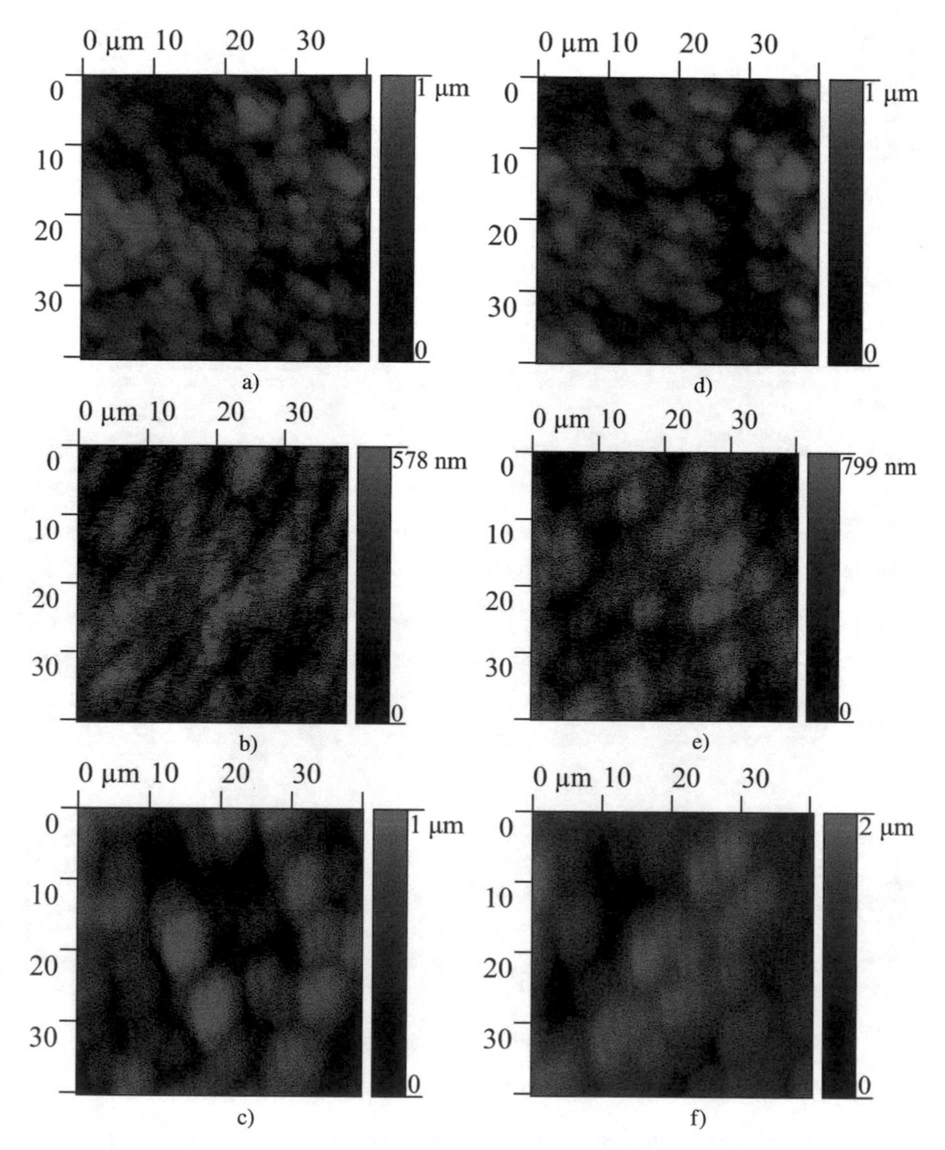

Fig. 8.4 AFM images of the surfaces after 45 seconds etching of the samples: a) 1, b) 2, c) 3 and after 60 seconds etching: d) 1, e) 2, f) 3

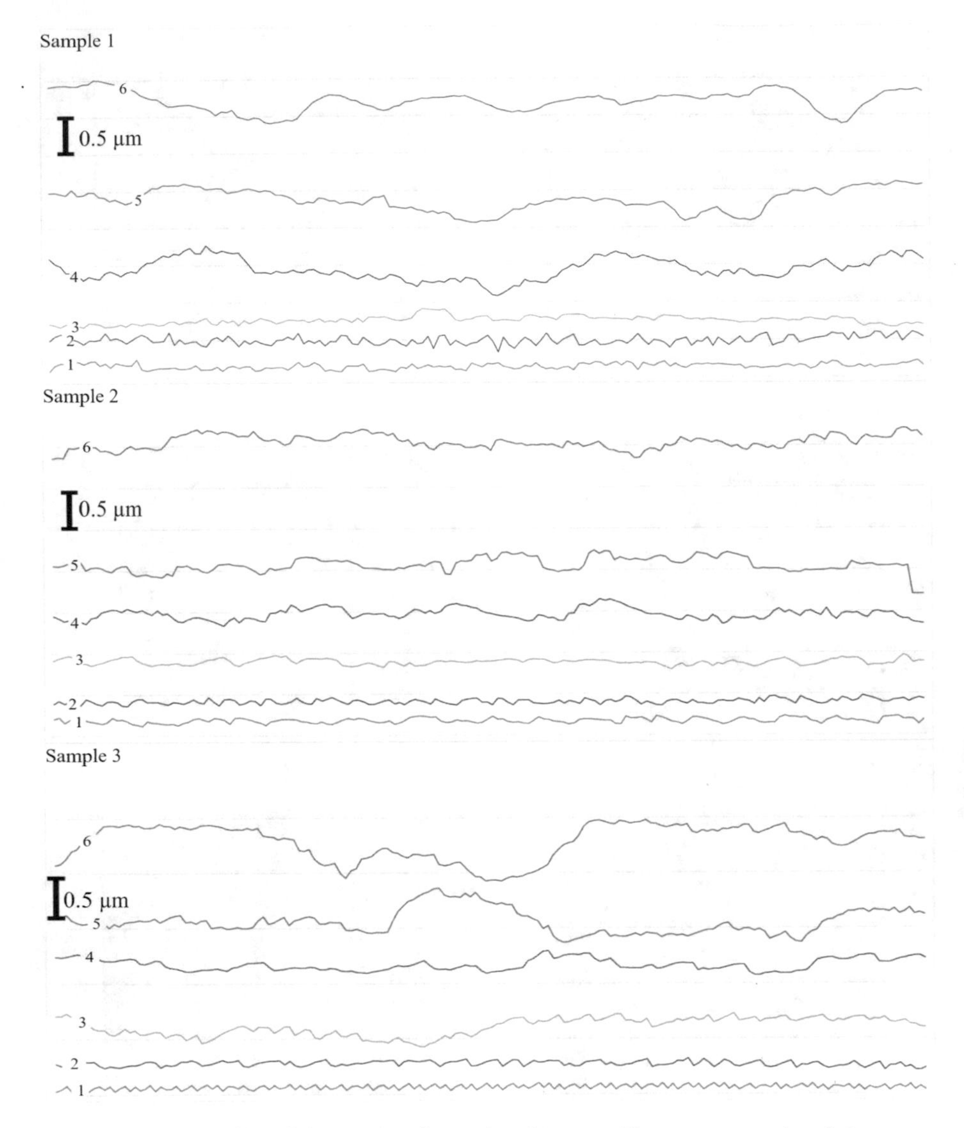

Fig. 8.5 Surface profiles of the samples after each etching step. The steps are numbered: 1 – without etching, 2 – etching for 1 s, 3 – etching for 15 s, 4 – etching after 30 s, 5 – etching after 45 s, 6 – etching after 60 s

featureless surface associated with the smooth smear layer generated during polishing. This layer was formed from the residues of the structural components of the enamel and rare inclusions of abrasive after the grinding/polishing procedures (Ostolopovskaya et al, 2013).

Examination of the surface of the samples after 15 s etching still showed the presence of the smear layer on samples 1 and 2 (Fig. 8.3a, b) which were etched

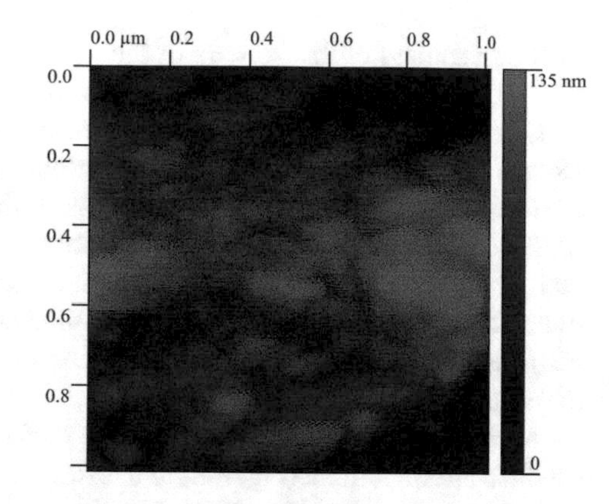

Fig. 8.6 High resolution image of sample 1 after 60 s etching

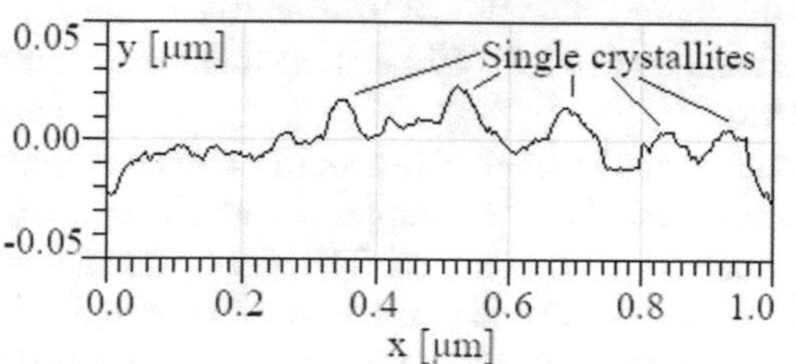

Fig. 8.7 Horizontal surface profile for the sample 1 after 60 s etching time obtained with high resolution

with the lower concentrations of citric acid. Irregular rounded structural elements – enamel prisms (Fig. 8.3c) are clearly visualized on the surface of sample 3. The latter observation suggests that fifteen seconds of etching with citric acid at a concentration of 5 wt. % was sufficient to completely remove the smear layer. A feature that is evident in the latter image is the differential dissolution rate of the enamel prisms. Surface monitoring after 30 s etching (Fig. 8.3d, e, f) on all samples demonstrated the presence of enamel prisms in the form of round and droplet-shaped elements. In addition the number of visualized structural elements increased. The more detailed structure is clearly seen on samples 2 and 3 (Fig. 8.4e, f). An additional feature that these and later images show is that there are different alignments of the rods that may be attributed to the different areas investigated of the same tooth. In samples 2 and 3 the rods appear to be inclined while in the sample 1 they are more normal to the surface.

Etching for 45 s resulted in acquiring of more pronounced outlines of the structural element components of samples 2 and 3 (Fig. 8.4b, c). The structure of sample 1 still remained poorly ordered. The non-etched inter-prism enamel remained as well on the surface of the samples 1 and 2 (Fig. 8.45a, 5b). After etching of the samples for 60 seconds no further significant changes in the structure of samples 1 and 3 were observed. (Fig. 8.4d, e). In the case of sample 1 acid concentration was still not enough to fully remove inter-prism enamel. The structural elements of sample 2 slightly increased in size, acquired more streamlined shape as the inter-prism enamel was etched. With increasing time of etching for all the samples an increase of the sharpness of the borders of the prisms were observed on the surface profiles (Fig. 8.5). However, this process is more pronounced for sample 3.

Comparing the results of etching the human enamel using 37% phosphoric acid (Retief and Denys, 1979) (where the authors obtained the mean surface roughness 260.9 nm after 30 s) the etching process in the present research seems slower. We managed to acquire the comparable value of roughness using 5 wt. % citric acid etching for 60 s. In this regard, we can conclude that citric acid of the mentioned concentration is more suitable for controlled etching than 37% phosphoric acid. It is also known that with prolonged exposure to phosphoric acid, enamel can be seriously damaged (Mirjanić et al, 2015).

According to the dependencies of the surface roughness Ra from the etching time on different directions (Fig. 8.8) and absolute value of the surface roughness Ra_{absol} from the etching time (Fig. 8.9) after 30 s etching the roughness increases in a rather stable manner, and after 45 s etching a linear increase in the values of Ra_{absol} can be observed.

Etching for 45 s resulted in acquiring of more pronounced outlines of the structural element components of samples 2 and 3 (Fig. 8.5b, c). The structure of sample 1 still remained poorly ordered. The non-etched inter-prism enamel remained as well on the surface of the samples 1 and 2 (Fig. 8.5a, b). After etching of the samples for 60 seconds no further significant changes in the structure of samples 1 and 3 were observed. (Fig. 8.5d, e). In the case of sample 1 acid concentration was still not enough to fully remove inter-prism enamel. The structural elements of sample 2 slightly increased in size, acquired more streamlined shape as the inter-prism enamel was etched. With increasing time of etching for all the samples an increase of the sharpness of the borders of the prisms were observed on the surface profiles (Fig. 8.6). However, this process is more pronounced for sample 3.

Comparing the results of etching the human enamel using 37% phosphoric acid (Retief and Denys, 1979, where the authors obtained the mean surface roughness 260.9 nm after 30 s) the etching process in the present research seems slower. We managed to acquire the comparable value of roughness using 5 wt. % citric acid etching for 60 s. In this regard, we can conclude that citric acid of the mentioned concentration is more suitable for controlled etching than 37% phosphoric acid. It is also known that with prolonged exposure to phosphoric acid, enamel can be seriously damaged (Mirjanić et al, 2015).

According to the dependencies of the surface roughness Ra from the etching time on different directions (Fig. 8.7) and absolute value of the surface roughness Ra_{absol} from the etching time (Fig. 8.9) after 30 s etching the roughness increases in a rather stable manner, and after 45 s etching a linear increase in the values of Ra_{absol} can be observed.

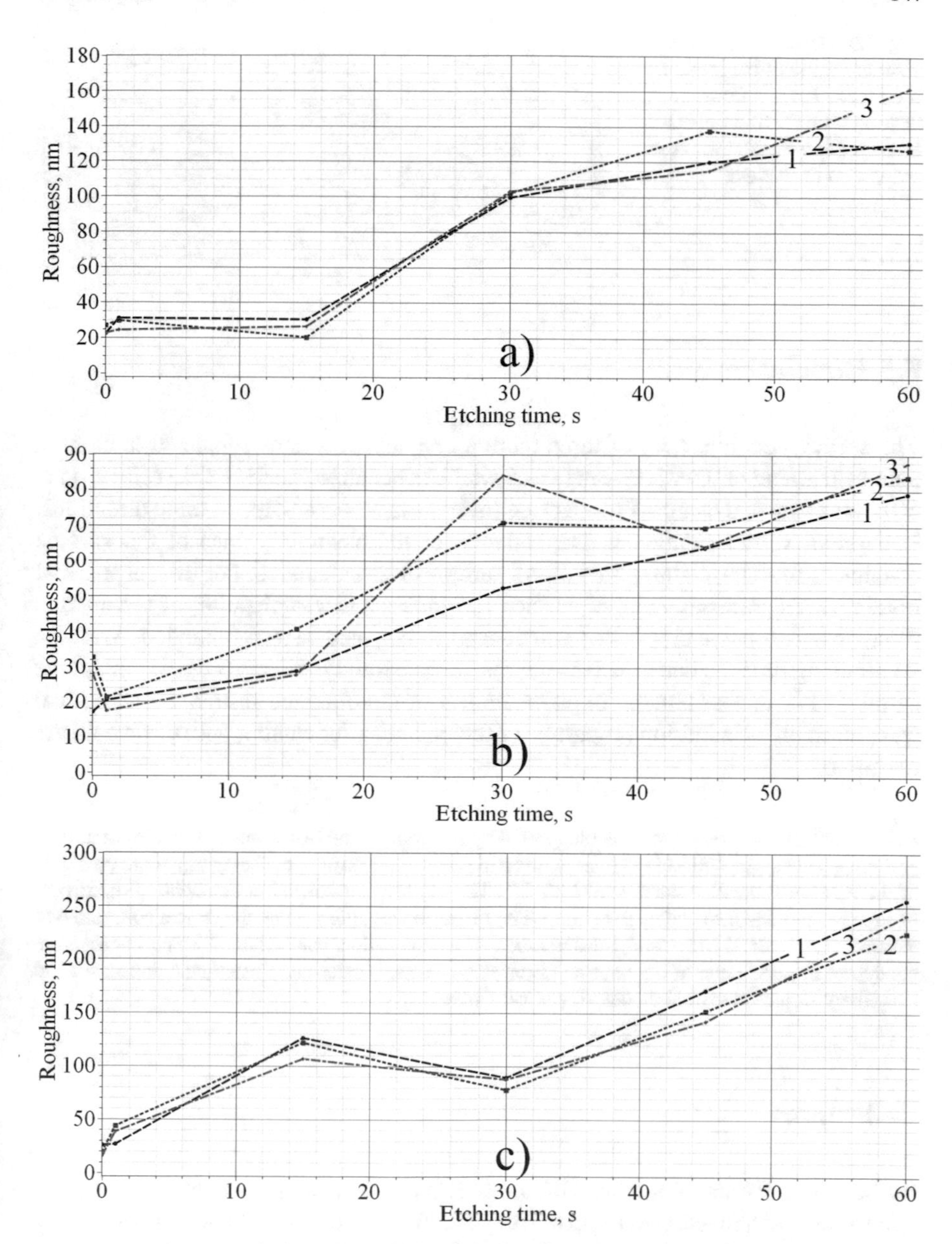

Fig. 8.8 Dependence of the surface roughness ***Ra*** from the etching time on different directions (1 – horizontal, 2 – vertical, 3 — diagonal) for: a) sample 1, b) sample 2, c) sample 3

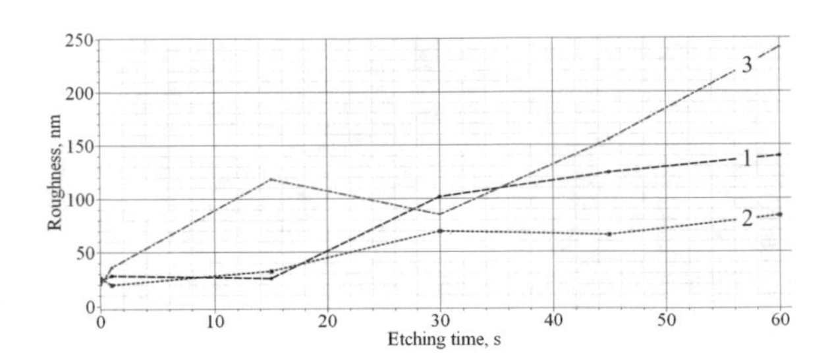

Fig. 8.9 Dependence of the absolute value of the surface roughness Ra_{absol} from the etching time for all samples. Each graph is marked with the number of the sample

8.5 Conclusion

The surface roughness *Ra* of three human enamel specimens subjected to etching using citric acid of 0.5 wt. %. 3 wt. %. 5 wt. % concentrations for 1 s, 15 s, 30 s, 45 s, 60 s was evaluated using AFM. Surface roughness was evaluated in three directions: horizontal, vertical, diagonal. Dependence of the absolute values of the surface roughness from the etching time for all samples was calculated. For the purposes of obtaining a significant value of surface roughness and visualization of the enamel prism in a reasonable time 5 wt. % citric acid etching can be recommended. After the 45 s etching time the linear increase of the surface roughness was observed on all the samples. For further clinical implementation of the obtained results the additional researches on the adhesion properties of the enamel after etching with citric acid are necessary.

Acknowledgements This work was supported by the grant of the Government of the Russian Federation (grant No. 14.Z50.31.0046). E. V. Sadyrin was supported by the scholarship of the President of the Russian Federation no. SP-3672.2018.1. The tooth was extracted in the Dental Department of Rostov State Medical University Clinic. The research was approved by the ethical committee of Rostov State Medical University, the agreement from the patient was obtained. Experiments were conducted in Nanocenter of Research and Education Center "Materials" (http://nano.donstu.ru) and Chemistry Department of Don State Technical University.

References

Barbour ME, Parker DM, Allen GC, Jandt KD (2003) Human enamel dissolution in citric acid as a function of pH in the range 2.30≤pH≤6.30 - a nanoindentation study. European Journal of Oral Sciences 111(3):258–262, DOI 10.1034/j.1600-0722.2003.00039.x

Cardenas AFM, Siqueira FSF, Bandeca MC, Costa SO, Lemos MVS, Feitora VP, Reis A, Loguercio AD, Gomes JC (2018) Impact of pH and application time of meta-phosphoric acid on resin-enamel and resin-dentin bonding. Journal of the Mechanical Behavior of Biomedical Materials 78:352 – 361, DOI 10.1016/j.jmbbm.2017.11.028

Dong Z, Chang J, Deng Y, Joiner A (2010) In vitro remineralization of acid-etched human enamel with Ca_3SiO_5. Applied Surface Science 256(8):2388 – 2391, DOI 10.1016/j.apsusc.2009.10.072

Erhardt MCG, Cavalcante LMA, Pimenta LAF (2004) Influence of phosphoric acid pretreatment on self-etching bond strengths. Journal of Esthetic and Restorative Dentistry 16(1):33–40, DOI 10.1111/j.1708-8240.2004.tb00449.x

Gordeeva NO, Egorova AV, Magomedov TB, Venatovskaya NV (2004) Methodology for reducing the risk of pathology of hard tissues of teeth with non-removable orthodontic treatment equipment. Saratov Journal of Medical Scientific Research 7(1):230–233

Landuyt KV, Kanumilli P, Munck JD, Peumans M, Lambrechts P, Meerbeek BV (2006) Bond strength of a mild self-etch adhesive with and without prior acid-etching. Journal of Dentistry 34(1):77 – 85, DOI 10.1016/j.jdent.2005.04.001

Lopes FM, Markarian RA, Sendyk CL, Duarte CP, Arana-Chavez VE (2006) Swine teeth as potential substitutes for in vitro studies in tooth adhesion: A sem observation. Archives of Oral Biology 51(7):548 – 551, DOI 10.1016/j.archoralbio.2006.01.009

Loyola-Rodriguez JP, Zavala-Alonso V, Reyes-Vela E, Patiño Marín N, Ruiz F, Anusavice K (2009) Enamel roughness and depth profile after phosphoric acid etching by atomic force microscopy. Journal of Electron Microscopy 59(2):119–125, DOI 10.1093/jmicro/dfp042

Mirjanić D, Mirjanić V, Vojinović J (2015) Testing the effect of aggressive beverage on the damage of enamel structure. Contemporary materials 1(6):55–61

Moura SK, Pelizzaro A, Bianco KD, De Goes MF, Loguercio AD, Reis A, Grande RHM (2006) Does the acidity of self-etching primers affect bond strength and surface morphology of enamel? The journal of adhesive dentistry 8(2):75–83

Ostolopovskaya OV, Anokhina AV, Ruvinskaya GR (2013) Modern adhesive systems in clinical dentistry. Practical Medicine 4(72):15–20

Retief DH, Denys FR (1979) Finishing of enamel surfaces after debonding of orthodontic attachments. The Angle Orthodontist 49(1):1–10, DOI 10.1043/0003-3219(1979)049<0001:FOESAD>2.0.CO;2

Sorozini M, dos Reis Perez C, Rocha GM (2018) Enamel sample preparation for afm: Influence on roughness and morphology. Microscopy Research and Technique 81(9):1071–1076, DOI 10.1002/jemt.23073

Torres-Gallegos I, Zavala-Alonso V, Patiño-Marín N, Martinez-Castañon GA, Anusavice K, Loyola-Rodríguez JP (2012) Enamel roughness and depth profile after phosphoric acid etching of healthy and fluorotic enamel. Australian Dental Journal 57(2):151–156, DOI 10.1093/jmicro/dfp042

van der Vyver PJ, de Wet F, van Rensburg JJM (1997) Bonding of composite resin using different enamel etchants. The Journal of the Dental Association of South Africa 52(3):169–172

Watari F (2005) In situ quantitative analysis of etching process of human teeth by atomic force microscopy. Journal of Electron Microscopy 54(3):299–308, DOI 10.1093/jmicro/dfi056

Zanet C, Arana V, Fava M (2006) Scanning electron microscopy evaluation of the effect of etching agents on human enamel surface. The Journal of Clinical Pediatric Dentistry 30(3):247–50, DOI 10.17796/jcpd.30.3.x5q47j208g064366

Zheng J, Xiao F, Qian L, Zhou Z (2009) Erosion behavior of human tooth enamel in citric acid solution. Tribology International 42(11):1558 – 1564, DOI 10.1016/j.triboint.2008.12.008, special Issue: 35th Leeds-Lyon Symposium

Chapter 9
Experimental and Numerical Methods to Analyse Deformation and Damage in Random Fibrous Networks

Emrah Sozumert, Emrah Demirci, and Vadim V. Silberschmidt

Abstract Deformation and damage behaviours of random fibrous networks are investigated with experimental and numerical methods at local (fibre) and global (specimen) levels. Nonwoven material was used as an example of fibrous network, with its individual fibres were extracted and tested with a universal testing system in order to assess their material properties. The fibres demonstrated a nonlinear time-dependent response to stretching. For analysis of notch sensitivity, undamaged nonwoven specimens and those with various notch shapes were analysed with fabric-level tensile tests and finite-element simulations. A level of strains around notch tips was tracked in simulations, demonstrating that the material was notch-sensitive, but load-transfer mechanisms were different than those in standard homogeneous materials. The notch shape also affected the rate of damage growth in the main directions. A good agreement between experimental and numerical damage patterns was observed. Also, the notch shape affected strength and toughness of the fibrous network.

Key words: Fibrous networks · Deformation · Damage · Finite-element modelling · Nonwovens

9.1 Introduction

This research focuses on mechanics of deformation and damage of random fibrous networks that are ubiquitous in natural environments (e.g. collagens) and synthetic materials, such as nonwovens. They can be produced by living microorganisms (such as bacteria) in aqueous environment (Gao et al, 2016a). They form a membrane in the

Emrah Sozumert · Emrah Demirci · Vadim V. Silberschmidt
Loughborough University, UK,
e-mail: e.sozumert@lboro.ac.uk, e.demirci@lboro.ac.uk,
v.silberschmidt@lboro.ac.uk

H. Altenbach et al. (eds.), *Plasticity, Damage and Fracture in Advanced Materials*, Advanced Structured Materials 121,
https://doi.org/10.1007/978-3-030-34851-9_9

form of amniotic sac to provide a protective environment for a developing embryo in human body (Buerzle and Mazza, 2013). Random fibrous networks can be produced with an electrospinning process and merging fibrous networks are used to host cells in a process of repairing skin injuries (Norouzi et al, 2015). Therefore, there is a growing interest in the improvement of existing fibrous networks and developing new ones thanks to their broad range of potential applications. This inevitably attracts the attention of researchers to the deformation and damage mechanisms of such networks. This paper presents a full methodology for mechanical characterisation of deformation and damage of a random fibrous network based on a combination of experimental and numerical approaches (from single-fibre to fabric tests).

9.1.1 Background

Random fibrous networks are discontinuous materials with their complex microstructures. Considering their deformation and damage behaviour, such networks have advantageous compared to traditional continuous materials, such as metals. For instance, Gao et al (2016b) and Ridruejo et al (2015) reported that a notch in a random fibrous network can improve material stiffness. Researchers in academia and industry investigate microstructures of these materials since microstructural parameters, such as alignments of fibres, influences their effective mechanical properties, such as the strength of the network, at macro-level (Sabuncuoglu et al, 2012). Additionally, these parameters can be tuned to enhance mechanical properties (Ovaska et al, 2017).

To study these materials, a nonwoven fibrous network with random orientation distribution of fibres is considered as an example material in this research. Fibres of this material underwent a thermal-bonding process and were bonded together by high pressure and temperature. Conditions of the web-bonding process, such as its temperature, controls mechanical properties of this type of random fibrous networks (Michielsen et al, 2006). In general, the manufacturing process results in anisotropy, with three main orthogonal directions: machine direction (MD), cross direction (CD), and thickness direction (TD). Fibres are preferentially aligned along MD as a result of manufactures, and, as tensile load applied, an anisotropic mechanical response is observed, with higher stiffness in MD than that in CD (Demirci et al, 2012).

The aim of this study is to present experimental and numerical methods used to investigate deformation and damage mechanisms of fibrous networks. In experimental methods, readers can find the techniques for characterisation of tensile behaviour of individual fibres as well as random fibrous networks. Fibres extracted from nonwoven networks were tested in a micro tester to assess their time-dependent elasto-plastic properties; while fabric tests were performed to observe deformation and damage evolutions at macro-scale. The data obtained from the single-fibre tests were used as input to numerical investigations (finite-element (FE) simulations). Scanning Electron Microscopy (SEM) and Micro Computed Tomography (micro CT) were employed to analyse microstructure of nonwoven samples and to provide data for development of parametric FE models. Microstructural features were directly incorporated into these models for samples without notches and with various notches, mimicking the microstructure of a selected random fibrous network. In order to evaluate the evolution of local deformation and fracture behaviour in fibrous networks, the FE models were stretched along their main direction - MD. Global and local (at fibre level) responses of the fibrous networks were recorded with strain distributions over central area and in the vicinity of notch tips tracked.

9.1.2 Fundamental Concepts

Complexity of deformation and damage mechanisms of random fibrous networks stems from their complex microstructure described in terms of orientation distribution of fibres, nonlinear behaviour of individual fibres and fibre curvature. The main source of nonlinearity related to orientation distribution of fibres is randomness. Two types of randomness can be encountered in a fibrous network:

(i) randomness in spatial alignment of fibres and
(ii) randomness in their orientation.

Such microstructure of fibrous networks dictates introduction of randomly distributed fibres in network domain. A fibre can be defined with a random orientation angle ϕ_1, and its centre in $x - y$ coordinate system C_1 (Fig. 9.1 (a)). The centre of this fibre can be placed in a different location C_2 (Fig. 9.1 (b)), to model the randomness in spatial alignments of fibres, or in the same location C_1 with a different orientation angle ϕ_3 (Fig. 9.1 (c)) (randomness in orientation of fibres).

Although the adjective *random* is used for fibrous networks, alignment of fibres in them are usually not fully random. In cases when the alignments of fibres are random, these random fibrous networks are called *fully random fibrous networks.*

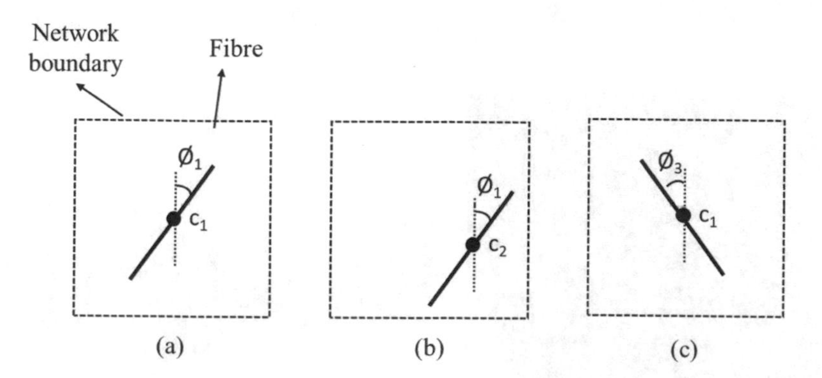

Fig. 9.1 Introduction of fibre with three different configurations: (a) reference configuration, (b) same orientation and different centre, (c) different orientation and same centre

Furthermore, orientation distribution of fibres is an important element, affecting mechanical properties of fibrous networks. In addition to random spatial distribution, some of these networks have preferential orientations of fibres, with some principal directions. It was reported that different stiffness and toughness measurements were obtained in the networks with a non-uniform orientation distribution of fibres (Yang et al, 2015). To express the randomness in orientation distribution of fibres, Cox (1952) introduced a distribution function for fibrous mats. This is also known as *orientation distribution function* (ODF). Determination of this function is important for elucidating actual microstructure of fibrous networks, as it is a fundamental input into numerical models. In nonwoven materials, there are three main directions defining orientation of fibres in corresponding material: MD (machine direction), CD (cross direction), TD (thickness direction). In fact, the machine direction is the direction coinciding with that of the production line and perpendicular to the cross direction. This orientation distribution of fibres can be assessed from SEM images by employing Fourier- or Hough-transform-based methods (Demirci et al, 2012; Ghassemieh et al, 2002; Kim and Pourdeyhimi, 2001). They are based on detection of fibres in fibrous networks. A sample SEM image and its computed ODF are presented in Fig. 9.2.

The third nonlinearity source is fibre curvature (in other words, fibre crimp or curliness), e.g. studied by Hearle and Stevenson (1963). A fibre with crimp shows negligible stiffness until it becomes fully straight. Overall mechanical behaviour of a fibrous network is directly affected by crimp of fibres (Kumar and Rawal, 2017; Shiffler, 1995). Crimp geometry was described by a sinusoidal-like wave (Rawal, 2006). Additionally, crimp can increase material ductility and toughness (Sozumert et al, 2018). Another source of nonlinearity in complex mechanical behaviour of random fibrous networks is a nonlinear material response of individual fibres (Farukh et al, 2012; Sabuncuoglu et al, 2013). Polymer fibres with diameters at micro-level were tested under tensile testing conditions in various investigations; their time-dependent material response was also obtained (Sabuncuoglu et al, 2013). Clearly, time-dependent material properties of fibres alter a tensile response of random fibrous

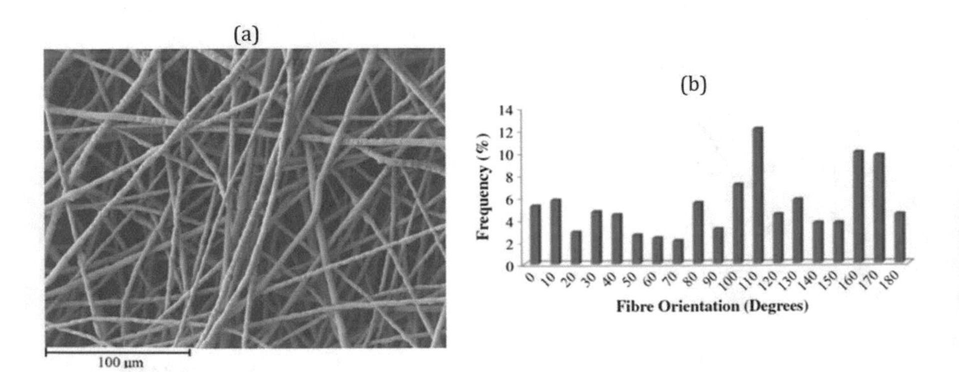

Fig. 9.2 SEM image of nonwoven (a) and its computed ODF (b) (Demirci et al, 2012)

networks. Testing of individual fibres is, however, not always feasible (particularly for fibres with diameters at nano-scale); especially when fibres cannot be easily removed from a fibrous network for testing. Hence, alternative ways, such as methods based on Atomic Force Microscopy (AFM) (Cheng and Wang, 2008; Tan et al, 2005) and/or numerical-based methods (Gao et al, 2017), can be used for characterisation of individual fibres. From the microscopic point of view, deformation and damage in random fibrous networks can be explained with few steps at micro-level (Farukh et al, 2014b; Ridruejo et al, 2011):

(i) debonding of fibre under low level of strain;
(ii) re-orientation of fibres towards loading direction;
(iii) straightening of fibres;
(iv) stretching and
(v) failure of individual fibres.

Microscopic appearance of a nonwoven felt subjected to axial stretch at maximum load is shown in Fig. 9.3 (a) and, as the load increased, damage occurred due to failure of fibre bonds followed by failure of fibres (Fig. 9.3 (b)) Ridruejo et al (2010). This shows damage starts at local level and spreads to global level. Since understanding of local damage and deformation micro-mechanisms of random fibrous networks is important for this research, inserting an artificial crack (i.e. notch) can ensure the localisation of damage around certain areas for a better and easier analysis. Therefore, the effect of local damage on structural integrity and material behaviour can be evaluated with notch sensitivity.

The finite-element (FE) method is used to simulate deformation and fracture of fibrous networks and to reveal information about fundamental micromechanisms underpinning these phenomena, such as in (Ghassemieh et al, 2002; Isaksson and Hägglund, 2009; Koh et al, 2013; Kulachenko and Uesaka, 2012; Mueller and Kochmann,

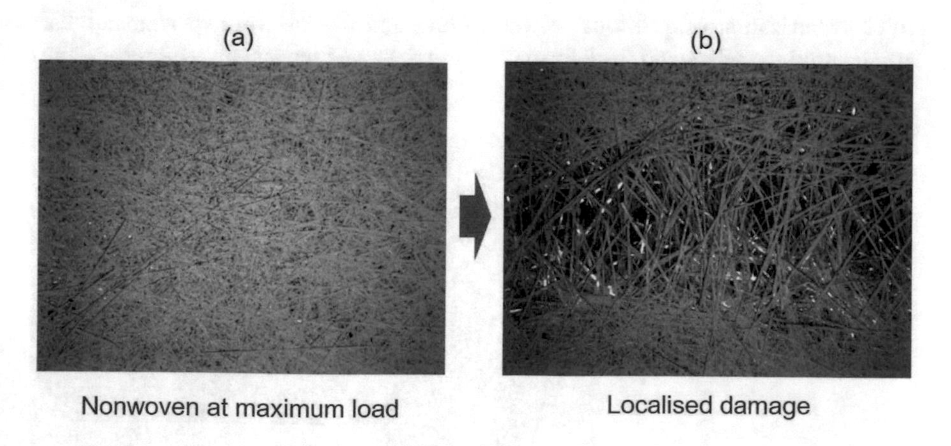

Fig. 9.3 Optical micrographs of nonwoven felt under tensile deformation at maximum load (a) and localised damage as stretched further (b) (modified from Ridruejo et al, 2010)

2004). Notch-sensitivity in random fibrous networks composed of brittle fibres is investigated in Isaksson and Hägglund (2009); Koh et al (2013); Ridruejo et al (2010) and, as for ductile fibres, in Ridruejo et al (2010). The fibrous networks with ductile fibres, such as some nonwovens, demonstrate a blunted notch area reducing stress-concentration. The fibrous networks with brittle fibres, such as gelatine electrospun scaffolds, however, protected the network morphology around the notch tip.

9.2 Experimentation

9.2.1 Material

A thermally bonded calendared 30 g/m^2 nonwoven composed of monocomponent polypropylene fibres is used as an example random fibrous network. To enhance mechanical properties by tuning crystallinity, after the propylene fibres were extruded, they were stretched (Jubera et al, 2014). This process was followed by lying down the spun fibres randomly on a moving conveyed belt. The fibres on the belt showed non-uniformity in their orientation distribution. At the subsequent stage, i.e. calendaring process, heat and pressure were simultaneously applied to bond fibres. This manufacturing process caused a clear distinction in bonded and unbonded regions of the network. The bonded fibres as bond points with a spatial pattern and unbonded bridging fibres as fibre matrix are shown in Fig. 9.4.

9.2.2 Experimental Procedure

In characterization of a fibrous network, fibre- and fabric-level experimental tests are required to understand its deformation and damage mechanisms. Fibres are fun-

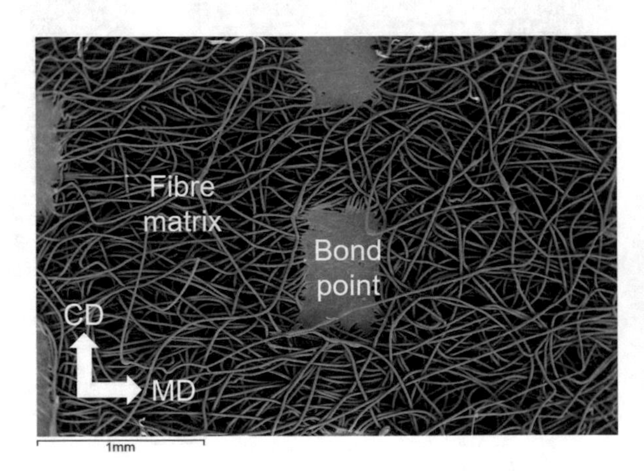

Fig. 9.4 SEM image of 30 g/m^2 point-bonded nonwoven

damental components of random fibrous networks, playing key roles in deformation mechanisms at micro-level. They, therefore, should be tested with tensile tests to quantify their material properties. Subsequently, the material properties are implemented in FE models of the fibrous networks. In addition to the fibre tests, nonwoven fabrics were tested in tension to verify the FE models and to investigate their notch sensitivity. Evolution of deformation and damage in experimental and numerical investigations were compared.

9.2.3 Single-fibre Tests

As discussed, material properties of individual fibres should be incorporated into FE models of random fibrous networks. One way to obtain these properties is to test the fibres under tensile conditions. It is known that physical properties of fibres are sensitive to manufacturing parameters, such as calendaring pressure and temperature. This means that their mechanical properties after the web-bonding process are different than those before the process. The data from tensile tests on processed and unprocessed fibres demonstrated significant differences, especially in their strengths (Mueller and Kochmann, 2004). Failure stress and corresponding strain values of processed and unprocessed fibres for various strain rates are shown in Table 9.1.

Employing a realistic and pragmatic approach, individual fibres extracted from nonwoven fabrics were used in a universal testing system to obtain strain-rate-dependent elastoplastic properties of fibres and to establish damage criteria for them in a FE model. Additionally, experimental observations pointed out that fibres of thermally point-bonded nonwovens failed at peripheries of these points. So, fibres were extracted from the nonwoven fabric with their bond points at the edges (see Fig. 9.5 (a)). This sample was placed between grips in the tensile tester, shown in Fig. 9.5 (b). To prevent any slippage between the fibre and the grips, sticky labels were attached to the bond points at the fibre edges. A schematic of the single-fibre specimen in grips is presented in Fig. 9.5 (b). The specimen was tested in Instron® Micro Tester 5848 with a high-precision 5 N load cell. A detailed procedure of this test was

Table 9.1 Failure stress and strain values of processed and unprocessed fibres at various strain rates (reproduced from Farukh et al, 2013)

Strain rate (1/s)	Fibre type	Failure stress (MPa)	Strain at failure stress
0.01	processed	481±48	1.29±0.67
	unprocessed	1299±95	2.93±0.78
0.1	processed	315±28	1.05±0.11
	unprocessed	998±57	1.77±0.09
0.5	processed	241±53	0.79±0.10
	unprocessed	818±60	1.49±0.14

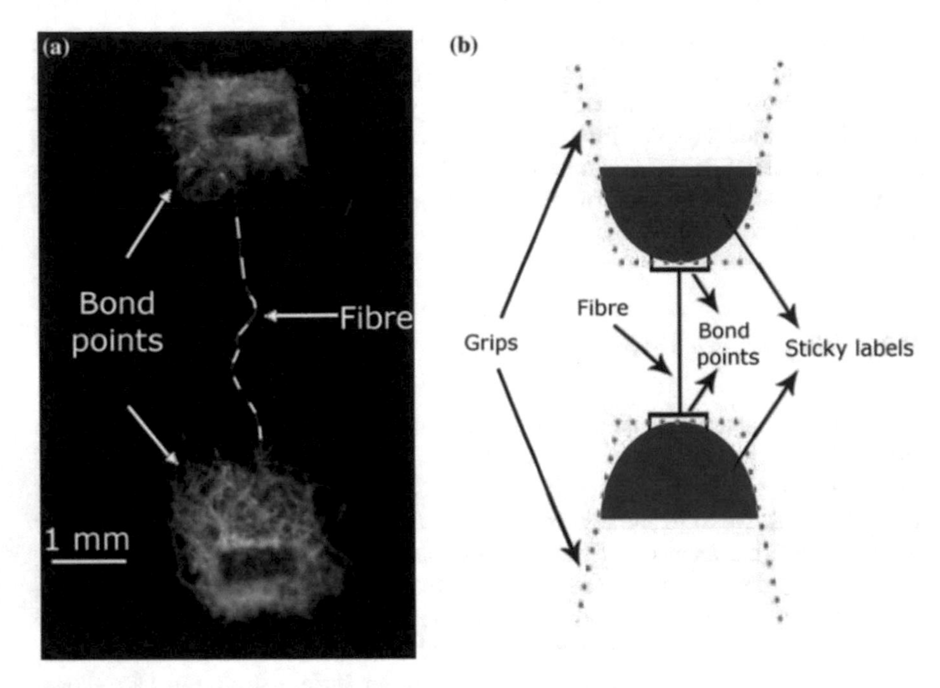

Fig. 9.5 (a) Single fibre with bond points at each end; (b) schematic of single-fibre specimen and its fixture Farukh et al (2013)

explained in Farukh et al (2013); Sabuncuoglu et al (2013). The tests were performed with a displacement control of cross-head movement, enabling a constant strain rate. The tensile tests were performed on processed PP fibres with various levels of strain rates - 0.01, 0.1, 0.5 1/s - and mean true stress-strain responses were calculated. Using a linear elasticity theory, the contribution of elastic strain was obtained and, then, removed from the true stress-strain curves, resulting in true stress-plastic strain curves (Fig. 9.6). For repeatability of the results, eight fibres for each strain rate were used. While calculating the stress-strain curves, an assumption of perfectly circular fibre cross-section was made, and the initial fibre diameter was 18 μm, measured from SEM images and justified with the manufacturer data. The curves in Fig. 9.6 were used as input into the FE models of random fibrous networks. Nonwoven-fabric tests verified the reliability of the FE models. Single-fibre tests provided the elastic modulus and the Poisson's ratio of fibres as 350 MPa and 0.42 Farukh et al (2014b). It was assumed that the material properties of bond points were the same.

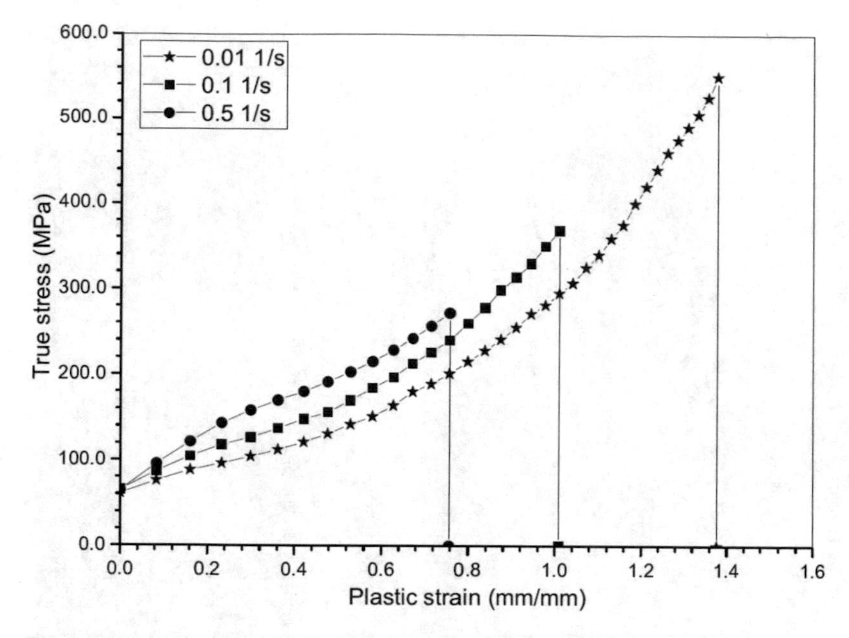

Fig. 9.6 Mean true stress-plastic strain curves for PP fibre at 0.01, 0.1, 0.5 1/s strain rates

9.2.4 Fabric Tests

Coupon specimens with square gauge area with the dimension of 25 x 25 mm^2 were prepared for fabric tests. For the notch-sensitivity analysis, four types of notches - circular, square, slit and diamond (see Fig. 9.7) - with the main dimension of 6.35 mm (D) were produced in the centres of the specimens using a surgery knife. To assess the effect of notches on deformation and damage behaviours of fibrous networks, non-damaged specimens were prepared and tested under tensile load. The specimens with notched and virgin geometries were tested in MD with a Hounsfield testing system (shown in Fig. 9.8) with a constant strain rate of 0.2 1/s. In the

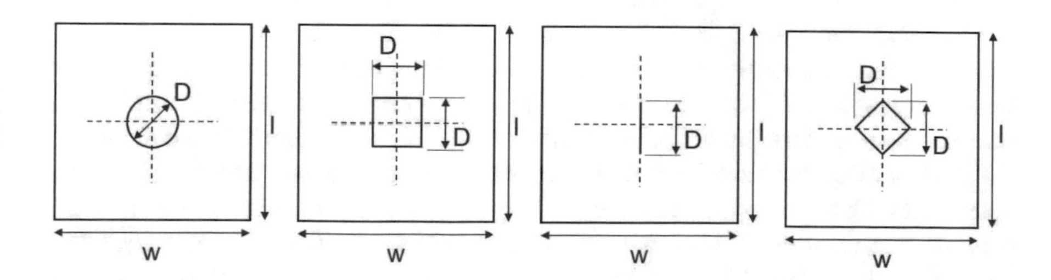

Fig. 9.7 Dimensions of gauge areas of coupon specimens with four different shapes of central notches: circular, square, slit and diamond

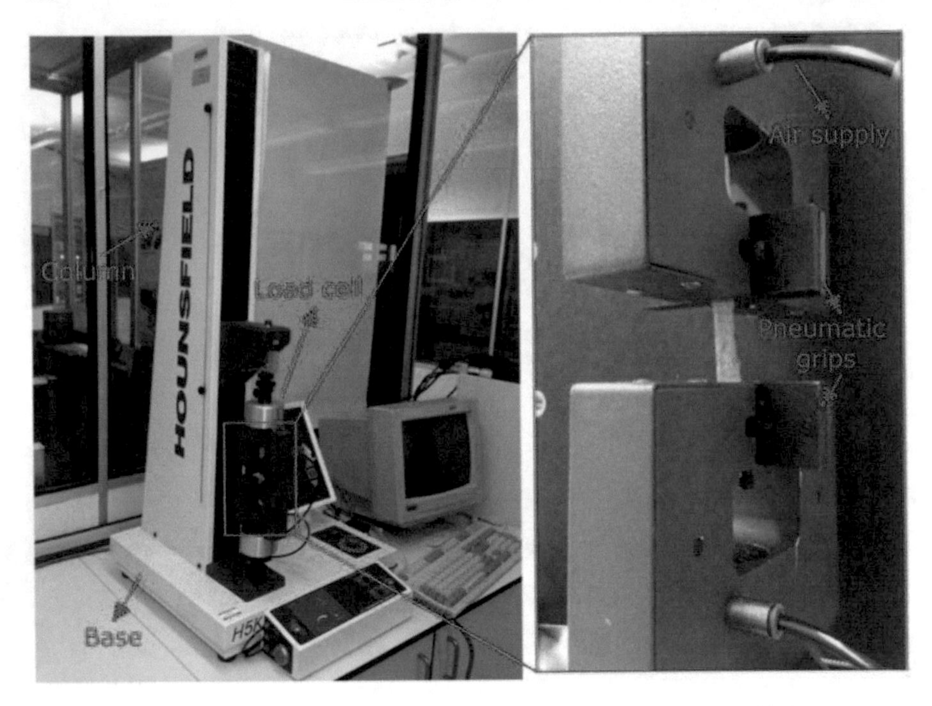

Fig. 9.8 Specimen in Hounsfield Benchtop Tester with pneumatic grips Farukh (2013)

experimental system, the specimens were fixed between pneumatic grips preventing any failure on clamped areas.

9.3 Numerical Investigations

For many years, researchers attempted to explain the mechanical behaviour of the fibrous networks using numerical models. In a pioneer work, H. L. Cox derived stiffness and strength of paper as a fibrous network in terms of orientation distribution of fibres and their elastic properties (Cox, 1952). M. A. Narter extended H. L. Cox's model for three-dimensional anisotropic fibrous networks (Narter et al, 1999). The finite-element method was one of the employed numerical methods; it is used in this research to simulate deformation and damage behaviours of random fibrous networks. FE models of such networks can be modelled with continuous and discontinuous approaches. Continuous models consider homogenisation of the networks to simulate macroscopic features of deformation; however, individual fibres are not generated and their failure caused by high strains cannot be calculated in continuous models. Conversely, discontinuous models aim to elucidate the effect of microstructure of the material, incorporating specific features, such as randomly distributed fibres, bond points etc. into FE simulations. In this way, underpinning deformation and damage processes can be explained under variety of loading conditions.

9.3.1 Discontinuous Finite-element Modelling of Random Fibrous Network

A micromechanical model was developed to obtain an insight into detailed features of deformation and damage of a random fibrous network. First, geometric features of microstructure of an undeformed network, such as dimensions of bond points and their pattern, were characterized with SEM; these data are available in (Sozumert et al, 2018). Additionally, the orientation distribution function (ODF) of fibres was computed with a Hough-transform-based algorithm (Demirci et al, 2012). The algorithm of the code, generating the FE model of a fibrous network, is illustrated in Fig. 9.9. Using this code, FE models of the coupon specimens with and without notches were generated in a commercial FE software, MSC Marc-Mentat. The notches were placed in the centre of specimen using x-y coordinate system and the same bond point pattern implemented for all notch shapes. The resultant FE model of the fibrous network is presented in Fig. 9.10 (a) with applied boundary conditions. In the FE software, fibres and bond points were modelled with two-node truss elements and thin-shell elements, respectively. Then, in a weak formulation for a momentum equation for a truss-shell system, the total internal virtual work δW_{int}, which is a sum of contributions from truss and shell elements, is equal to the total external virtual work δW_{ext}:

$$\delta W_{\text{int}} = \delta W_{\text{ext}},$$
$$\delta W_{\text{int}} = \sum_{i=1}^{n} \delta W_{\text{int}}^{\text{truss}} + \sum_{i=1}^{m} \delta W_{\text{int}}^{\text{shell}}. \tag{9.1}$$

The internal virtual work for a truss - in absence of transverse shear and moments - and for a shell element based on a Discrete Kirchhoff Theory in absence of shear deformation in x and y (in-plane) are given by

$$\delta W_{\text{int}}^{\text{truss}} = \int_0^L (N_z \delta \varepsilon_z)\, dz,$$
$$\delta W_{\text{int}}^{\text{shell}} = \int_0^L \left(M_x \delta \kappa_z + 2M_{xy} \delta \kappa_{xy} + M_y \delta \kappa_y\right) dA. \tag{9.2}$$

The virtual work of a truss is defined by the work done by axial force N_z, where, in a local coordinate system of a truss, z-axis coincides with the longitudinal axis. Additionally, the virtual work of a shell is the sum of a virtual work done by the

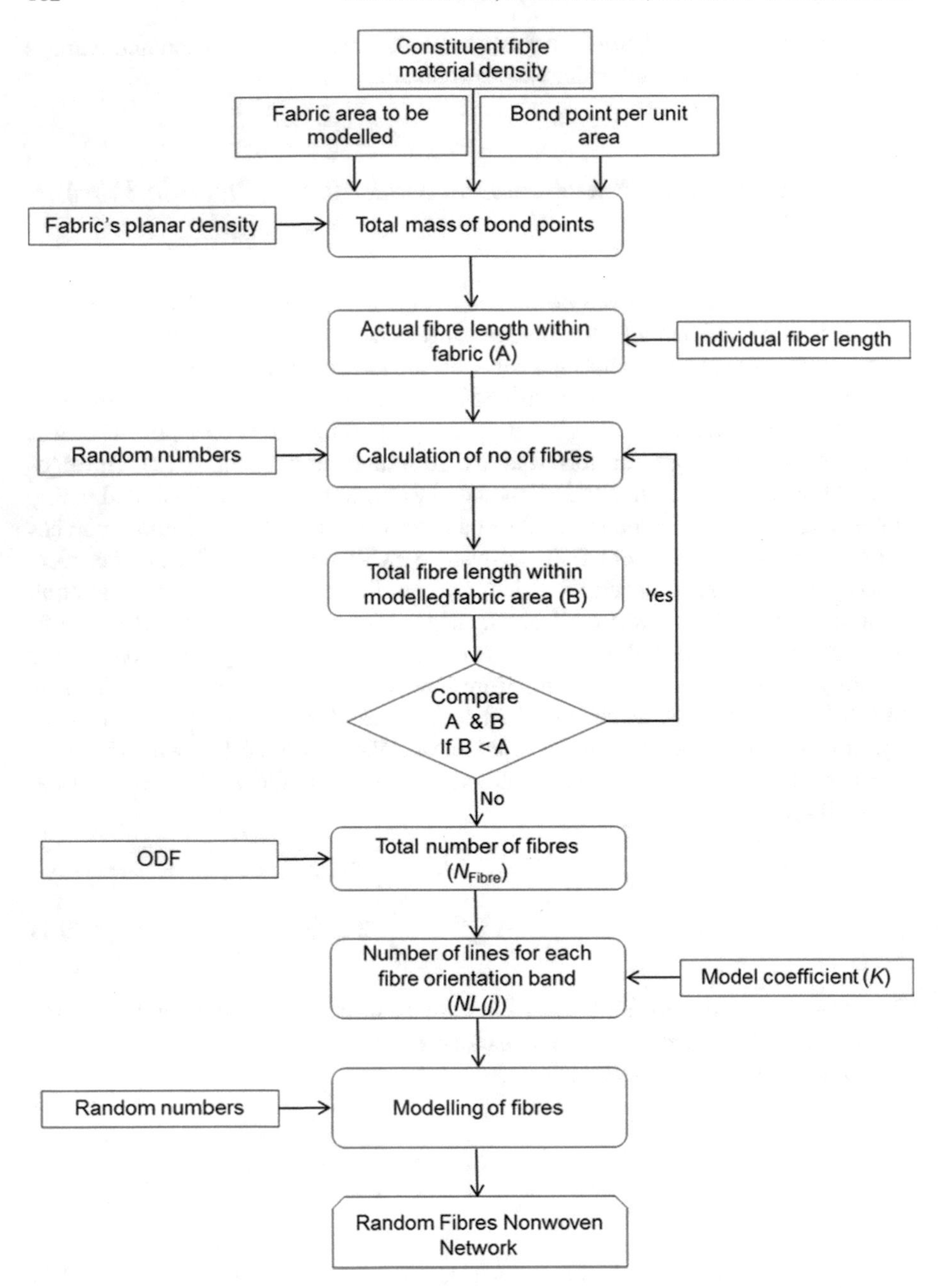

Fig. 9.9 Algorithm to generate a random fibrous network Farukh et al (2014a)

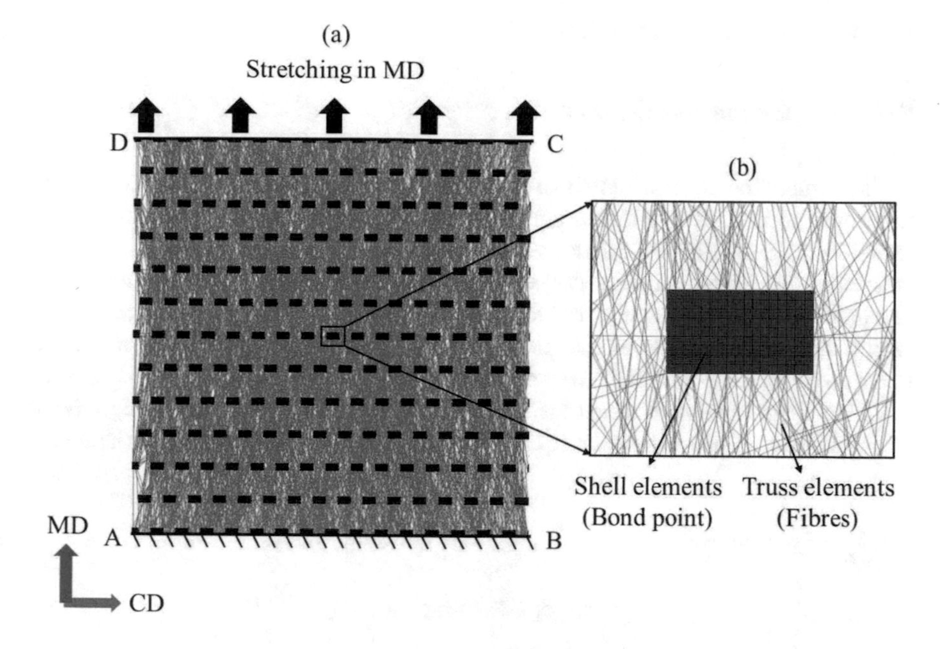

Fig. 9.10 (a) FE model with boundary conditions; (b) zoomed-in view of bond point and fibres

bending moments M_x and M_y, and the twisting moment M_{xy}, where κ_y, κ_z and κ_{xy} are curvatures. In the FE model, the bond points were discretised by thin-shell elements due to their relatively low thickness compared to their length and width (Fig. 9.4). This analysis considered that the bond points and fibres were represented one-to-one in the discrete FE models. Together with this, material behaviour of single fibres was incorporated in the incremental form of elastic-plastic stress-strain relations (MSC-Software, 2016; Belytschko et al, 2013). further information about the constitutive relations can be found in Sozumert et al (2018).

Some fundamental assumptions of the research should be mentioned: fibre curvature and interactions were neglected. For tensile loading, the curvature might however increase the toughness of the material by controlling the maximum stress and strain.

The simulations of tensile tests were performed with the FE models with virgin and notched configurations by applying a set of boundary conditions to the lower boundary AB and the upper boundary CD (see Fig. 9.10 (a)):

(i) the nodes on AB were fully fixed, reflecting the exact boundary conditions of pneumatic grips in the experimental tests;
(ii) the nodes on CD were moved with a uniform axial displacement in MD.

It is noted that FE simulations were carried out with an implicit solver under quasi-static loading conditions by using a large-deformation formulation.

9.4 Results and Discussion

9.4.1 Experimental Observations

To investigate the effect of notch shape on deformation and damage behaviour, the nonwoven specimens with a virgin microstructure and various notch geometries were tested in machine direction under tensile loading and force-displacement readings were obtained from the Instron® system. To make a comparison between the virgin and notched cases, the respective force curves were normalised by using the effective width (the difference between the specimen's width and the notch width). The (mean) normalised force-extension curves of these specimens in MD are shown in Fig. 9.11 (a). Five specimens were tested for each geometric configuration and a significant scatter in force-displacement data after the onset of damage was observed; therefore,

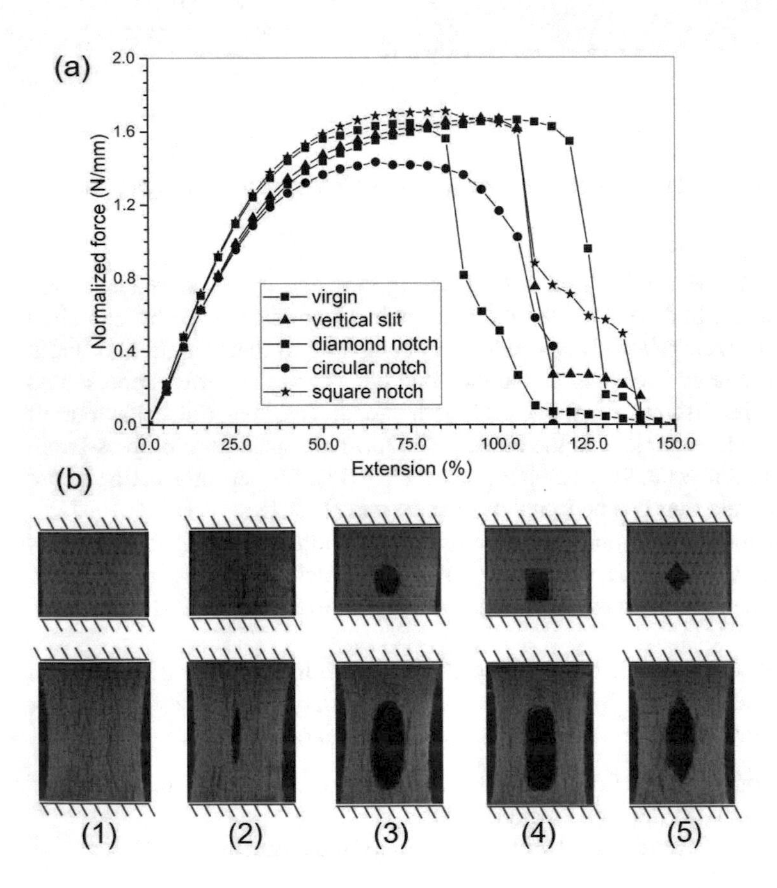

Fig. 9.11 (a) Experimental mean force-extension curves for virgin and notched specimens tested in MD. (b) Damage evolution in various nonwoven specimens under 60% extension in MD: (1) virgin; (2) vertical slit; (3) circular notch; (4) square notch; (5) diamond notch (Sozumert et al, 2018)

mean curves were plotted. Based on Fig. 9.11 (a), deformation behaviour of the fibrous network was sensitive to the notch shape, but to different extent. Mean tensile strength of specimens, however, was insensitive to the vertical slit. The underlying reason for this is the fact that the ODF of the selected fibrous network showed a preferential orientation of fibres in MD as compared to CD. As the vertical slit notch was opened, the amount of fibres cut, making contribution to the load-carrying capacity, was negligible.

Damage evolution of the nonwoven specimens with virgin and damaged configurations is shown in Fig. 9.11 (b). Several deformation processes related to constituting fibres of the random fibrous network take place in microstructure:

(i) fibre straightening from initial curved form;
(ii) realigning of fibres towards the loading direction;
(iii) fibre stretch and failure.

From the perspective of macrostructure, at the early stage of tensile stretch, the central notches grew in both MD and CD. After the realignment of fibres was completed, the speed of damage growth in MD was much faster than the one in CD.

Microstructural observations of virgin specimens of thermally-point bonded nonwovens under tensile load revealed that, after straightening of fibres, the fibres were rotated towards the loading direction and their progressive failure followed this process when their failure strain was reached (Farukh et al, 2014b). Experiments showed that these deformation and damage mechanisms were independent of the basis weight, and the strain rate of the tensile test did not change the sequence of deformation and damage phenomena (Ridruejo et al, 2011).

The changes in the microstructure of the virgin nonwoven sample at the different level of extensions are showed in Fig. 9.12. Fibres were bundled around bond points and weak areas at arbitrary locations, where lower density of fibres was observed. These weak areas, i.e. zones of damage localisations, expanded with further deformation, with fibres failing in the areas supporting this expansion (Sozumert et al, 2018). Arrows in Fig. 9.12 indicate fibres failed as a result of large strain. For the specimens with square, diamond and circular notches, the position of damage-localisation areas was more predictable as compared to virgin specimens: they were near the notches. Re-orientation of fibres towards the loading direction toughened the material, while sharp edges of the notches were blunted. Bond points in various locations of the fibrous network were rotated by different amounts as a result of a non-uniform distribution of individual fibres in the network. Deformation and load transfer mechanisms between bond points and their attached fibres are highly unpredictable since the load was transferred via random sets of links between bond points and fibres (i.e. discontinuous load-transfer mechanism). Therefore, local loads were diffused rather than concentrated with stress singularity near a notch as in continuous materials (such as metals).

The numerical results (Fig. 9.11 (a)) provided an evidence that the levels of strength of specimens with virgin and notches were almost the same. Apparently, the specimens with circular holes showed poorer tensile performance than the rest of the specimens. A limited investigation on circular- and longitudinal-notched specimens

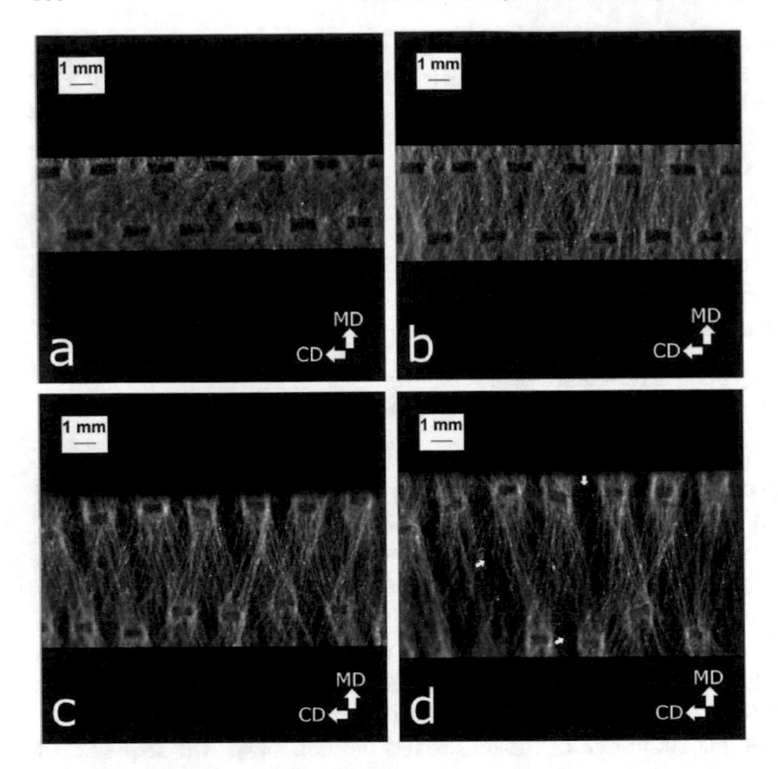

Fig. 9.12 Deformation and damage mechanisms of virgin nonwoven specimen: (a) 0% extension; (b) 25% extension; (c) 50% extension; (d) 80% extension (Farukh et al, 2014b)

was conducted in Rawal et al (2013), revealing the similar detrimental effect of circular notches on the tensile response of needle-punched hybrid nonwovens. One hypothesis is that a higher number of fibres contributing to the load-carrying capacity was affected as compared to other notch cases.

9.4.2 Finite-element Simulations

In the FE simulations, the true stress-true strain curve (0.01 1/s in Fig. 9.6) was implemented into an elastic-plastic material model with von Mises plasticity in FE software. This curve showed a softer mechanical response than that used in simulations in Sozumert et al (2018). Throughout this investigation, the strain rate effects on deformation and damage behaviour were not considered. A maximum-strain damage criterion for fibres was implemented into the FE models of the nonwoven specimens that were stretched until they failed completely. The damage parameter D and the failure criteria are then expressed as

$$\text{if } D = \left(\frac{\varepsilon_z}{\varepsilon_f}\right) \geq 1.0, \text{ fibre fails.} \tag{9.3}$$

where ε_z and ε_f are the longitudinal logarithmic strain (simulation) and the mean logarithmic failure strain (experiment), respectively. In simulations, mono-component fibres were assumed to show a fully isotropic material behaviour, and the failure stress $\varepsilon_f = 1.4$ was taken from single-fibre tests. In the FE models, each truss elements represented a fibre segment and, as the longitudinal strain ε_z in it exceeded the failure stress ε_f, it was assumed that the fibre (or truss element) failed and it was deleted.

Tensile behaviour of nonwoven specimens with virgin, vertical-slit, square and circular notches was successfully simulated with the developed model and the assigned damage criterion for fibres. Microstructural changes at different levels of extensions were analysed. Strains of fibre segments were obtained from the deformed FE models of virgin nonwoven specimens and normalised by the failure strain of fibres. A histogram of the normalised fibre strains at 40%, 60%, 80%, and 100% fabric extensions is presented in Fig. 9.13. More than 90% of fibre segments shows normalised strains in the range between 0-0.4 in 40% extension, far below the amount required for failure. As the FE model was stretched further, almost 10% of fibre segments approached to the value of 1.0. A further increase in the extension value resulted in failure of constituent fibres in the virgin specimen, leading to a change of the network's microstructure. It was therefore observed that some of fibres in the loading direction (MD) shifted back to their original position due to local stress

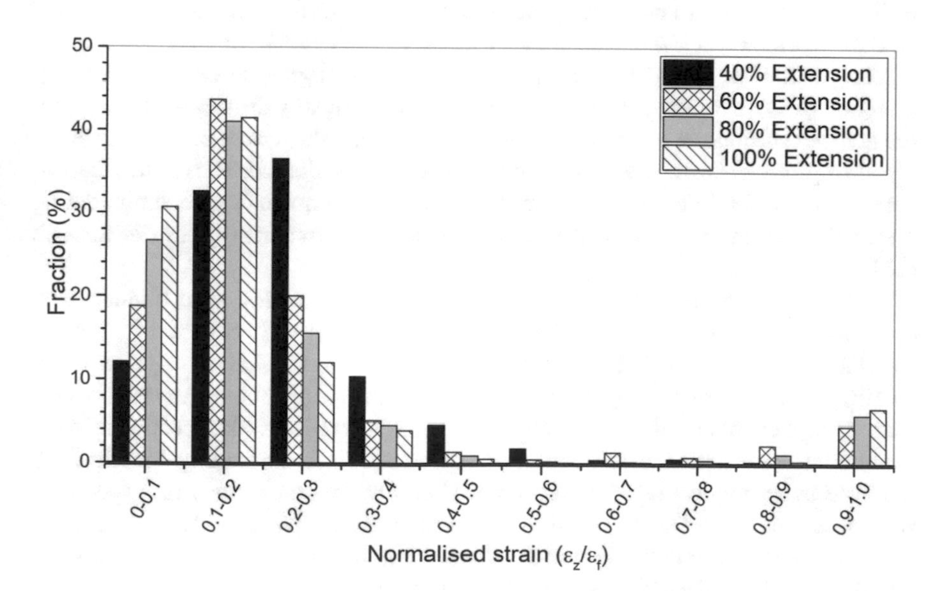

Fig. 9.13 Distributions of fibre strains (normalised by failure strain of fibres) in virgin specimen (simulation) for fabric extension of 40%, 60%, 80%, and 100%

release. However, as the network was further stretched, those fibres were re-aligned in MD.

Images of damage patterns for the virgin, slit-, circular- and square-notched specimens for 0%and 60% extensions were obtained from experimental and numerical tensile tests (Fig. 9.14). The FE models with various geometries were capable to

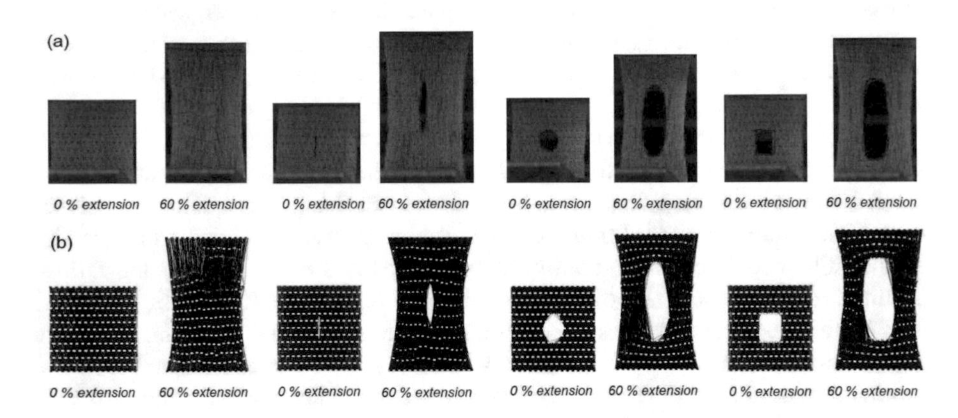

Fig. 9.14 Comparison of experimental (a) and simulated damage (b) patterns for virgin, slit-notch, circular-notch and square-notch specimens for 0% and 60% extensions

demonstrate the main deformation and damage features, such as damage growth in orthogonal directions and necking behaviour. Fibres in the fibrous network were randomly oriented at 0% extension. As they were subjected to uniaxial stretch, the notches in the specimens opened and fibres were re-aligned towards the loading direction (MD). The amount of re-alignment was not equally distributed throughout the notched specimens; therefore, failures of individual fibres were more localised in certain regions where the fibres were more preferentially aligned. The FE simulations demonstrated that fibres bundles were formed in front of notch tips, accommodating mechanical strain and resisting crack propagation, thus toughening the fabric (Koh et al, 2013).

In the experiments, sharp edges in the notched specimens were blunted, and this geometric variation due to fabric deformation was well captured in the FE simulations. The square and slit notches became a rectangle-like damage area and an elliptical hole, respectively. Like the slit notch, the circular hole was transformed into an elliptical hole, but with a relatively lower aspect ratio. As the specimens in the simulations were stretched, re-alignments of fibres towards the loading direction was followed by re-alignments of bond points. Their distribution in the virgin specimen was quite similar to that in the slit-notch specimen at 30% extension, where the aspect ratio of ellipse in the slit notch increased. Alignments of bond points around the square and circular notches were almost the same.

In the FE models, fibres along selected paths were tracked to assess the localisation of strain; Fig. 9.15 shows these paths for virgin, slit-notch, square-notch and circular-notch specimens. They were chosen in the centre of specimens, parallel to the loading

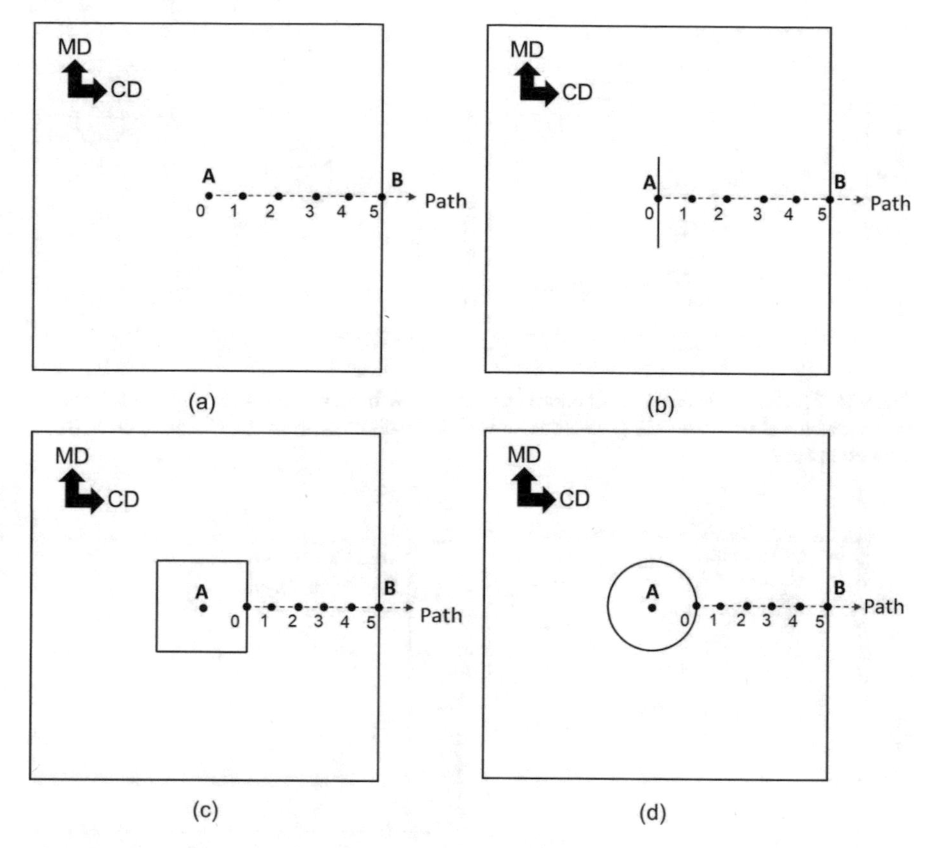

Fig. 9.15 Paths selected for specimens with virgin (a), slit-notch (b), square-notch (c) and circular-notch (d) geometries

direction and divided into five equal intervals. Logarithmic axial strains ε_z of fibres in these intervals were calculated for 60% (Fig. 16 (a)) and 80% (Fig. 16 (b)) stretching of the networks and averaged for each interval.

Non-uniform strain distributions were obtained for the tracked paths. Apparently, the normalised strains increased towards the notch tips (node-0) in Fig. 9.16 (a) and (b), demonstrating notch-sensitivity. Additionally, the highest normalised (averaged) strains were observed at the tip of slit notch. Due to lateral contraction and fibre alignment, the virgin specimen showed a nonlinear behaviour along the path, but lower strains as compared to the notched specimens. In terms of location (with the same $x - y$ coordinates, red-circled in Fig. 9.16 (a) and (b)), all cases of notches demonstrated similar damage performance. The slit-notched specimen, however, resulted in a sharper increase in the vicinity of the notch tip than the other notch cases, i.e. more localised strain around it. Unlike the slit-notched specimen, the square-notch one dispersed strain more out of the notch region.

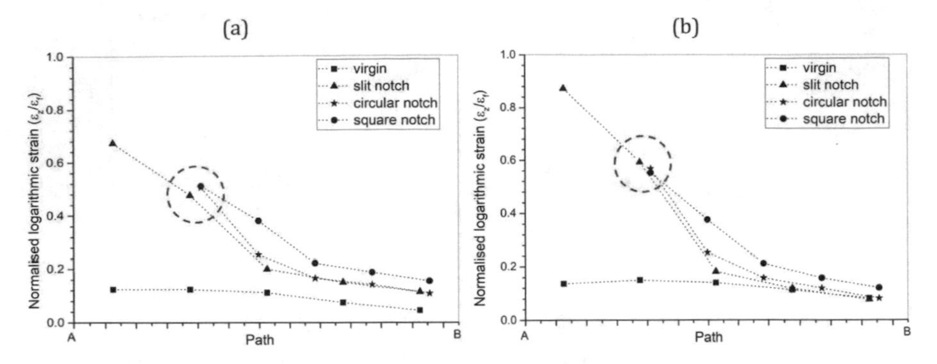

Fig. 9.16 FE-based calculated distributions of fibre strains for specimens with virgin, slit-notch, square-notch and circular-notch geometries for various levels of network stretching: (a) 60%; (b) 80% extensions

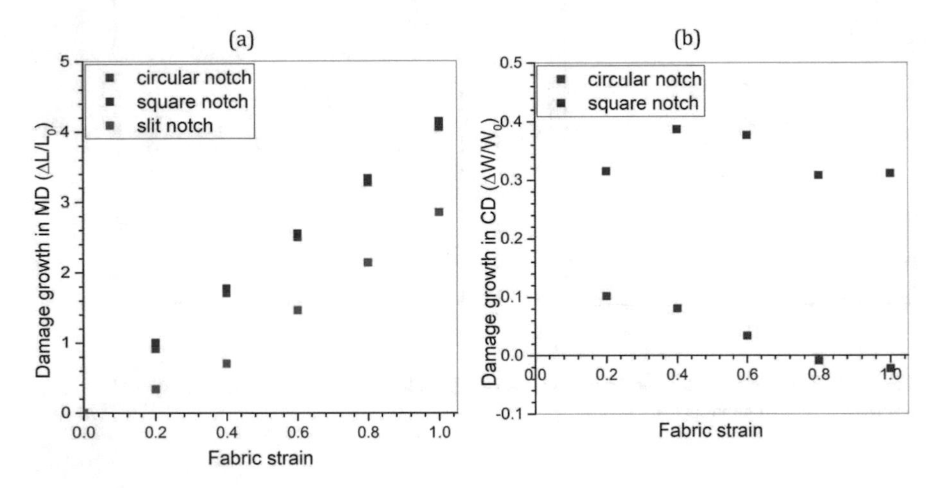

Fig. 9.17 Damage growth in MD (a) and CD (b)

With application of digital image processing of calculated damage patterns of the notched specimens, changes in damage length ΔL (MD) and width ΔW (CD) for various levels of fabric stretching were measured and normalised by initial length L_0 and width W_0, respectively. The damage growth in MD and in CD are demonstrated in Fig. 9.17. For all the notch cases, the damage growth in MD was almost linearly proportional to fabric strain. For damage growth in MD for the square- and circular-notched specimens showed similar behaviour and higher than that of the slit-notch specimen. As for the growth in CD, square- and circular-notched specimens were tracked and a nonlinear response was observed (Fig. 9.17 (b)). Opposite to the behaviour in MD, the square notch grew larger in CD than the circular notch. It increased at the initial stage, till 0.4 of fabric strain and started shrinking. Similarly, the circular notch shrank after a certain extent of stretching, followed by a decrease, even somehow below the initial notch width W_0.

9.5 Summary and Conclusions

This study explained experimental and numerical methods for characterizing deformation and damage behaviour of random fibrous networks, with a special focus on deformation and damage evolution from local (fibre) to global (fabric) levels in a nonwoven material. Tensile tests were conducted on polymer fibres extracted from the nonwoven in order to obtain their material properties. The fibres demonstrated a nonlinear time-dependent material response to stretching. Undamaged nonwoven samples and ones with various notches were experimentally and numerically analysed. This was accompanied by development of discrete FE models imitating a real material microstructure. The FE models reproduced the deformation and damage mechanisms of fibrous networks at micro- and macro-scales. It was demonstrated that the notch shape had a significant effect on toughness of the network, with the circular notch having the most detrimental effect as compared to slit, square and diamond notches. In the FE simulations, using image processing, damage growth was tracked in two main orthogonal directions (MD and CD) as the specimens were stretched in MD. It was found that the rate of damage growth differed in various directions. This behaviour is linked to the main deformation and damage mechanisms of the fibrous network: rearrangement, stretching and failures of fibres.

References

Belytschko T, Liu WK, Moran B, Elkhodary K (2013) onlinear Finite Elements for Continua and Structures. Wiley

Buerzle W, Mazza E (2013) On the deformation behavior of human amnion. Journal of Biomechanics 46(11):1777 – 1783, DOI 10.1016/j.jbiomech.2013.05.018

Cheng Q, Wang S (2008) A method for testing the elastic modulus of single cellulose fibrils via atomic force microscopy. Composites Part A: Applied Science and Manufacturing 39(12):1838 – 1843, DOI 10.1016/j.compositesa.2008.09.007

Cox HL (1952) The elasticity and strength of paper and other fibrous materials. British Journal of Applied Physics 3(3):72–79, DOI 10.1088/0508-3443/3/3/302

Demirci E, Acar M, Pourdeyhimi B, Silberschmidt VV (2012) Computation of mechanical anisotropy in thermally bonded bicomponent fibre nonwovens. Computational Materials Science 52(1):157 – 163, DOI 10.1016/j.commatsci.2011.01.033

Farukh F (2013) Experimental and numerical analysis of deformation and damage in thermally bonded nonwoven material. PhD thesis, Loughborough University

Farukh F, Demirci E, Acar M, Pourdeyhimi B, Silberschmidt VV (2012) Strength of fibres in low-density thermally bonded nonwovens: An experimental investigation. Journal of Physics: Conference Series 382:012,018, DOI 10.1088/1742-6596/382/1/012018

Farukh F, Demirci E, Acar M, Pourdeyhimi B, Silberschmidt VV (2013) Meso-scale deformation and damage in thermally bonded nonwovens. Journal of Materials Science 48(6):2334–2345, DOI 10.1007/s10853-012-7013-y

Farukh F, Demirci E, Acar M, Pourdeyhimi B, Silberschmidt VV (2014a) Large deformation of thermally bonded random fibrous networks: microstructural changes and damage. Journal of Materials Science 49(11):4081–4092, DOI 10.1007/s10853-014-8100-z

Farukh F, Demirci E, Sabuncuoglu B, Acar M, Pourdeyhimi B, Silberschmidt VV (2014b) Numerical analysis of progressive damage in nonwoven fibrous networks under tension. International Journal of Solids and Structures 51(9):1670 – 1685, DOI 10.1016/j.ijsolstr.2014.01.015

Gao X, Shi Z, Kusśmierczyk P, Liu C, Yang G, Sevostianov I, Silberschmidt VV (2016a) Time-dependent rheological behaviour of bacterial cellulose hydrogel. Materials Science and Engineering: C 58:153 – 159, DOI 10.1016/j.msec.2015.08.019

Gao X, Shi Z, Liu C, Yang G, Silberschmidt VV (2016b) Fracture behaviour of bacterial cellulose hydrogel: Microstructural effect. Procedia Structural Integrity 2:1237 – 1243, DOI 10.1016/j.prostr.2016.06.158

Gao X, Sozumert E, Shi Z, Yang G, Silberschmidt VV (2017) Assessing stiffness of nanofibres in bacterial cellulose hydrogels: Numerical-experimental framework. Materials Science and Engineering: C 77:9 – 18, DOI 10.1016/j.msec.2017.03.231

Ghassemieh E, Acar M, Versteeg H (2002) Microstructural analysis of non-woven fabrics using scanning electron microscopy and image processing. Part 1: Development and verification of the methods. Proceedings of the Institution of Mechanical Engineers, Part L: Journal of Materials: Design and Applications 216(3):199–207, DOI 10.1177/146442070221600305

Hearle JWS, Stevenson PJ (1963) Nonwoven fabric studies: Part III: The anisotropy of nonwoven fabrics. Textile Research Journal 33(11):877–888, DOI 10.1177/004051756303301101

Isaksson P, Hägglund R (2009) Strain energy distribution in a crack-tip region in random fiber networks. International Journal of Fracture 156:1–9, DOI 10.1007/s10704-009-9340-9

Jubera R, Ridruejo A, González C, LLorca J (2014) Mechanical behavior and deformation micromechanisms of polypropylene nonwoven fabrics as a function of temperature and strain rate. Mechanics of Materials 74:14 – 25, DOI 10.1016/j.mechmat.2014.03.007

Kim HS, Pourdeyhimi B (2001) Computational modeling of mechanical performance in thermally point bonded nonwovens. Journal of Textile and Apparel, Technology and Management 1(4):1–7

Koh C, Strange D, Tonsomboon K, Oyen M (2013) Failure mechanisms in fibrous scaffolds. Acta Biomaterialia 9(7):7326 – 7334, DOI 10.1016/j.actbio.2013.02.046

Kulachenko A, Uesaka T (2012) Direct simulations of fiber network deformation and failure. Mechanics of Materials 51:1 – 14, DOI 10.1016/j.mechmat.2012.03.010

Kumar V, Rawal A (2017) Elastic moduli of electrospun mats: Importance of fiber curvature and specimen dimensions. Journal of the Mechanical Behavior of Biomedical Materials 72:6 – 13, DOI 10.1016/j.jmbbm.2017.04.013

Michielsen S, Pourdeyhimi B, Desai P (2006) Review of thermally point-bonded nonwovens: Materials, processes, and properties. Journal of Applied Polymer Science 99(5):2489–2496, DOI 10.1002/app.22858

MSC-Software (2016) Theory and user information

Mueller DH, Kochmann M (2004) Numerical modeling of thermobonded nonwovens. International Nonwovens Journal 13(1):56–62, DOI 10.1177/1558925004os-1300114

Narter MA, Batra SK, Buchanan DR (1999) Micromechanics of three-dimensional fibrewebs: constitutive equations. Proceedings of the Royal Society of London Series A: Mathematical, Physical and Engineering Sciences 455(1989):3543–3563, DOI 10.1098/rspa.1999.0465

Norouzi M, Boroujeni SM, Omidvarkordshouli N, Soleimani M (2015) Advances in skin regeneration: Application of electrospun scaffolds. Advanced Healthcare Materials 4(8):1114–1133, DOI 10.1002/adhm.201500001

Ovaska M, Bertalan Z, Miksic A, Sugni M, Benedetto CD, Ferrario C, Leggio L, Guidetti L, Alava MJ, Porta CAML, Zapperi S (2017) Deformation and fracture of echinoderm collagen networks. Journal of the Mechanical Behavior of Biomedical Materials 65:42 – 52, DOI 10.1016/j.jmbbm.2016.07.035

Rawal A (2006) A modified micromechanical model for the prediction of tensile behavior of nonwoven structures. Journal of Industrial Textiles 36(2):133–149, DOI 10.1177/1528083706067691

Rawal A, Patel SK, Kumar V, Saraswat H, Sayeed MA (2013) Damage analysis and notch sensitivity of hybrid needlepunched nonwoven materials. Textile Research Journal 83(11):1103–1112, DOI 10.1177/0040517512467063

Ridruejo A, González C, LLorca J (2010) Damage micromechanisms and notch sensitivity of glass-fiber non-woven felts: An experimental and numerical study. Journal of the Mechanics and Physics of Solids 58(10):1628 – 1645, DOI 10.1016/j.jmps.2010.07.005

Ridruejo A, González C, LLorca J (2011) Micromechanisms of deformation and fracture of polypropylene nonwoven fabrics. International Journal of Solids and Structures 48(1):153 – 162, DOI 10.1016/j.ijsolstr.2010.09.013

Ridruejo A, Jubera R, González C, LLorca J (2015) Inverse notch sensitivity: Cracks can make nonwoven fabrics stronger. Journal of the Mechanics and Physics of Solids 77:61 – 69, DOI 10.1016/j.jmps.2015.01.004

Sabuncuoglu B, Acar M, Silberschmidt VV (2012) Finite element modelling of thermally bonded nonwovens: Effect of manufacturing parameters on tensile stiffness. Computational Materials Science 64:192 – 197, DOI 10.1016/j.commatsci.2012.02.043

Sabuncuoglu B, Demirci E, Acar M, Silberschmidt VV (2013) Analysis of rate-dependent tensile properties of polypropylene fibres used in thermally bonded nonwovens. The Journal of The Textile Institute 104(9):965–971, DOI 10.1080/00405000.2013.766391

Shiffler DA (1995) An examination of the stress-strain curve of crimped polyethylene terephthalate staple fibers. The Journal of The Textile Institute 86(1):1–9, DOI 10.1080/00405009508631305

Sozumert E, Farukh F, Sabuncuoglu B, Demirci E, Acar M, Pourdeyhimi B, Silberschmidt VV (2018) Deformation and damage of random fibrous networks. International Journal of Solids and Structures DOI 10.1016/j.ijsolstr.2018.12.012

Tan EPS, Goh CN, Sow CH, Lim CT (2005) Tensile test of a single nanofiber using an atomic force microscope tip. Applied Physics Letters 86(7):073,115, DOI 10.1063/1.1862337

Yang W, Sherman VR, Gludovatz B, Schaible E, Stewart P, Ritchie R, Meyers MA (2015) On the tear resistance of skin. Nature Communications 6(6649), DOI 10.1038/ncomms7649

Chapter 10
Modelling of Ductile Fracture of Strain-hardening Cement-based Composites - Novel Approaches Based on Microplane and Phase-field Method

Christian Steinke, Imadeddin Zreid, and Michael Kaliske

Abstract Strain-hardening cement-based composites (SHCCs) denote a class of composite materials, which consist of a finely grained cementitious matrix and short, randomly orientated polymer micro fibers. At tensile loading, an initial linear-elastic stage results in the formation of a first crack in the matrix, that is bridged by micro fibers and exhibit a very limited crack mouth opening. Additional tensile loading reveals further increase in the load-bearing capacity at relatively large strains and a high energy absorption capacity due to the formation of additional micro cracks. The ultimate load is governed by structural failure due to pull out or rupture of the fibers bridging a critical crack. A novel approach models the composite behavior by a plasticity formulation with isotropic hardening, that is coupled to a numerical approximation for damage or ductile fracture to account for pull out or rupture of fibers, respectively.

10.1 Introduction

The prediction of the structural behavior of civil engineering constructions at excessive loading, e.g. earthquake or impact, is a wide field of ongoing research. Sophisticated numerical models are developed and applied for the simulation of plasticity, damage and crack evolution in order to identify vulnerabilities of the construction. The findings can be considered during the design of new buildings and allow for an efficient reinforcement of existing structures.

Strain-hardening cement-based composites (SHCCs) denote a class of recently developed composite material. A finely grained cementitious matrix is reinforced by polymer micro fibers. In analogy to standard steel bar reinforced concrete, the

Christian Steinke · Imadeddin Zreid · Michael Kaliske
Institute for Structural Analysis, TU Dresden, Germany,
e-mail: Christian.Steinke@tu-dresden.de, Imadeddin.Zreid@tu-dresden.de, Michael.Kaliske@tu-dresden.de

H. Altenbach et al. (eds.), *Plasticity, Damage and Fracture in Advanced Materials*, Advanced Structured Materials 121,
https://doi.org/10.1007/978-3-030-34851-9_10

excellent characteristics of the cement matrix at compressive loading is combined with the outstanding performance of polymer fibers at tensile loading. The formation of cracks in the matrix activates the fibers, that bridge the crack and maintain the structural integrity. Increased tensile loading can be applied and leads to the development of additional micro cracks bridged by fibers. The process is accompanied by high absorption of energy at relatively large deformations. The final structural failure is governed by pull out or rupture of fibers. The aggregate size in SHCC is restricted to grains of sand and the fibers are short. Therefore, the composite can be applied as a thin layer on existing structures. Due to the high energy absorption capacity at tensile loading, SHCC is especially useful for reinforcement of tensile sections of plates and beams loaded in bending, see e.g. Fig. 10.1.

A novel concept to approximate the characteristic behavior of SHCC at tensile loading is proposed. In general, the evolution of multiple micro cracks, that is accompanied by large deformation and energy dissipation, is modeled by a plasticity formulation with isotropic hardening. Structural failure, governed by pull out or rupture of fibers, is approximated by a gradient enhanced microplane model with damage and plasticity or a phase-field model for ductile fracture, respectively. The second section contains the theoretical framework of both approaches. Their application to a tension test on a dumbbell specimen is presented in the third section. Another numerical example demonstrates and discusses the approximation of a realistic post-fracture behavior with the phase-field approach. Finally, conclusions and an outlook closes the paper.

10.2 Modeling Approaches for Ductile Fracture of Strain-hardening Cement-based Composites

Cracks due to tension in the matrix of SHCC are bridged by the polymer fibers directly after the crack's formation. Therefore, the tensile load bearing capacity remains intact on the structural level. Crack opening is accompanied by a drop in the reaction force. Further increase of tensile loading results in additional micro cracks. This also involves relatively large deformations and a high amount of energy dissipation. A schematic diagram of this behavior and its representation by a numerical model are shown in Fig. 10.2b. Experimental evidence has been presented in Curosu (2017),

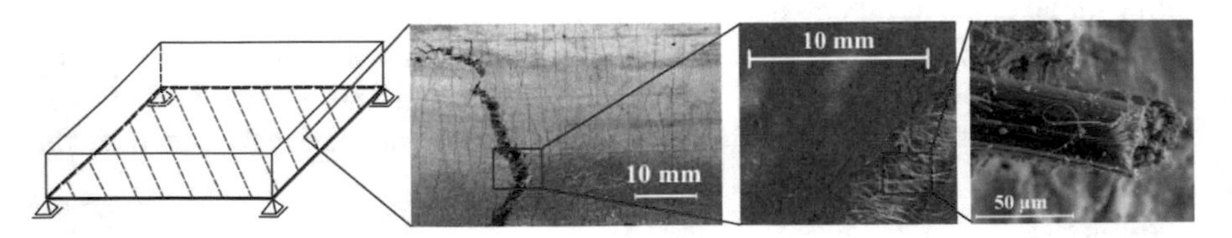

Fig. 10.1 Schematic view of the reinforcement of the tensile section of a plate by SHCC and zoom in on a micro crack bridged by fibers (adapted from Curosu, 2017)

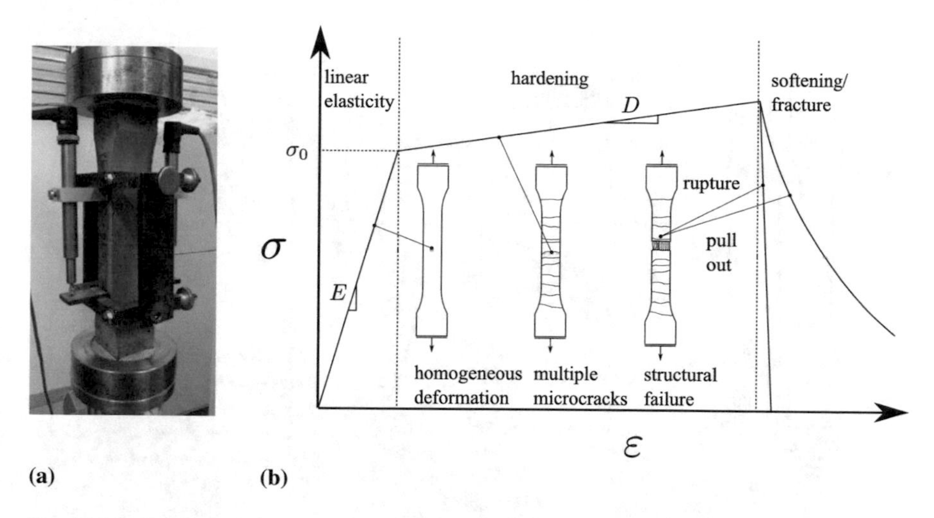

Fig. 10.2 Experimental investigation of SHCC dumbbell specimen at tensile loading for low strain rates adapted from Curosu (2017): (a) Experimental setup and (b) typical relation between mean stress and strain at the structural level and its representation by the numerical model

see e.g. Figs. 10.3b and 10.3c for the results. Unloading leads to a significant amount of permanent deformation. At the structural level, this behavior is similar to the characteristic of plastic material with hardening, i.e. the first micro crack occurs at the yield limit and the increased load bearing capacity at large strains is approximated by linear hardening.

Final structural failure is governed by pull out or rupture of fibers in a critical micro crack. Here, the material of the polymer fibers has an influence on the dominating mechanism. During pull out, additional frictional effects result in a relatively slow, softening-like degradation up to total failure, see e.g. Fig. 10.3b. In principle, this can be modeled by a damage approach. In the following, a gradient enhanced microplane model with damage and plasticity is proposed for this purpose. Rupture of the fibers results in a more abrupt failure, see Fig. 10.3c. In general, such a behavior can be approximated by fracture. Here, a phase-field model for ductile fracture is proposed.

10.2.1 Gradient Enhanced Microplane Model with Damage and Plasticity

The model for the gradient enhanced microplane approach with damage and plasticity, that is applied to the SHCC behavior in this study, is a well established approach for the simulation of reinforced concrete. The theoretical framework, its implementation and relevant information for its practical application within the scope of reinforced concrete modeling are outlined in detail in Zreid and Kaliske (2018). In

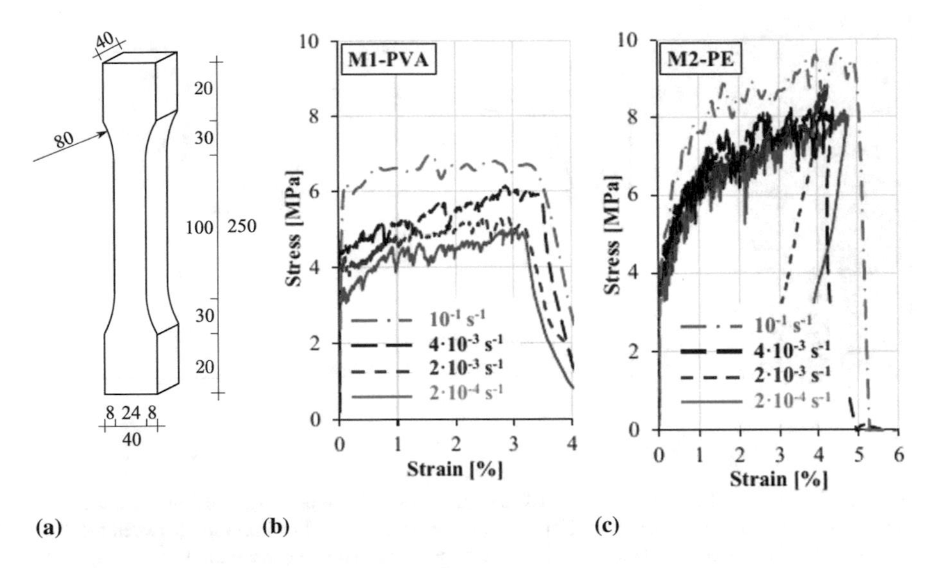

Fig. 10.3 (a) Geometry of the dumbbell specimen, relation between stresses and strains for reinforcement by (b) PVA (polyvinyl-alcohol) and (c) PE (polyethylene) based on experiments in Curosu (2017)

the following, the basic aspects of the model are recapitulated and classified with respect to its application for the SHCC characteristics.

The constitutive relation between strains and stresses for the microplane approach is established by an intermediate step, that involves a projection of the macroscopic strains to the set of microplanes. For the model at hand, a set of 42 microplanes and a volumetric-deviatoric split at the microplane level are applied. The projection of the macroscopic strain $\boldsymbol{\varepsilon}$ to the volumetric component $\varepsilon_{\mathrm{V}}^{\mathrm{mic}}$ and the deviatoric part $\boldsymbol{\varepsilon}_{\mathrm{D}}^{\mathrm{mic}}$ at a specific microplane is based on the vector $\mathbf{n}$ normal to that microplane and reads

$$\varepsilon_{\mathrm{V}}^{\mathrm{mic}} = \mathbf{V} : \boldsymbol{\varepsilon} \qquad \text{and} \qquad \boldsymbol{\varepsilon}_{\mathrm{D}}^{\mathrm{mic}} = \mathbf{n} \cdot \mathbf{Dev} : \boldsymbol{\varepsilon}\,, \tag{10.1}$$

with the projection tensors

$$\mathbf{V} = \frac{1}{3}\mathbf{1} \qquad \text{and} \qquad \mathbf{Dev} = \mathbf{I}^{\mathrm{sym}} - \frac{1}{3}\mathbf{1} \otimes \mathbf{1}, \tag{10.2}$$

where $\mathbf{1}$ is the second order identity tensor and $\mathbf{I}^{\mathrm{sym}}$ is the symmetric fourth order identity tensor. Then, a constitutive relation relates the microplane strain components $\varepsilon_{\mathrm{V}}^{\mathrm{mic}}$ and $\boldsymbol{\varepsilon}_{\mathrm{D}}^{\mathrm{mic}}$ to microplane stress components $\sigma_{\mathrm{V}}^{\mathrm{mic}}$ and $\boldsymbol{\sigma}_{\mathrm{D}}^{\mathrm{mic}}$, that are homogenized with respect to the microplane damage variable d^{mic} in order to yield the macroscopic stress tensor

$$\boldsymbol{\sigma} = \frac{3}{4\pi}\int_{\Omega}\left(1 - d^{\text{mic}}\right)\left(\sigma_{\text{V}}^{\text{mic}}\mathbf{V} + \boldsymbol{\sigma}_{\text{D}}^{\text{mic}}\right)d\Omega \qquad (10.3)$$
$$= \sum_{\text{mic}=1}^{21}\left(1 - d^{\text{mic}}\right)\left(\sigma_{\text{V}}^{\text{mic}}\mathbf{V} + \boldsymbol{\sigma}_{\text{D}}^{\text{mic}}\right)w^{\text{mic}}$$

by application of reduced numerical integration with the integration point weights w^{mic}. The constitutive equations at the microplane level read

$$\sigma_{\text{V}}^{\text{mic}} = K^{\text{mic}}\left(\varepsilon_{\text{V}}^{\text{mic}} - \varepsilon_{\text{V}}^{\text{mic,pl}}\right) \quad \text{and} \quad \boldsymbol{\sigma}_{\text{D}}^{\text{mic}} = 2G^{\text{mic}}\left(\mathbf{n}\cdot\mathbf{Dev}\right)^{\text{T}}\cdot\left(\varepsilon_{\text{V}}^{\text{mic}} - \varepsilon_{\text{V}}^{\text{mic,pl}}\right), \qquad (10.4)$$

where $\bullet^{\text{pl}}$ denotes a plastic component and K^{mic} and G^{mic} are the bulk and shear microplane moduli, respectively.

The smooth three-surface microplane cap yield function, as shown in Fig. 10.4, is defined by

$$f^{\text{mic}} = \frac{3}{2}\boldsymbol{\sigma}_{\text{D}}^{\text{mic}} : \boldsymbol{\sigma}_{\text{D}}^{\text{mic}} - f_{\text{DP}}^{2} f_c f_t. \qquad (10.5)$$

The DRUCKER-PRAGER yield function with hardening

$$f_{\text{DP}} = \sigma_0 - \alpha\sigma_{\text{V}}^{\text{mic}} + f_h(\kappa), \qquad (10.6)$$

considering an initial yield stress σ_0 and the friction coefficient α, is multiplied with a tension cap

$$f_t = 1 - H\left(\sigma_{V}^{\text{mic}} - \sigma_{V}^{T}\right)\frac{\left(\sigma_{V}^{\text{mic}} - \sigma_{V}^{T}\right)^2}{\left(T - \sigma_{V}^{T}\right)^2}, \qquad (10.7)$$

where

$$T = T_0 + R_t f_h(\kappa), \qquad (10.8)$$

and a compression cap

$$f_c = 1 - H\left(\sigma_{V}^{C} - \sigma_{V}^{\text{mic}}\right)\frac{\left(\sigma_{V}^{\text{mic}} - \sigma_{V}^{C}\right)^2}{X^2}, \qquad (10.9)$$

where

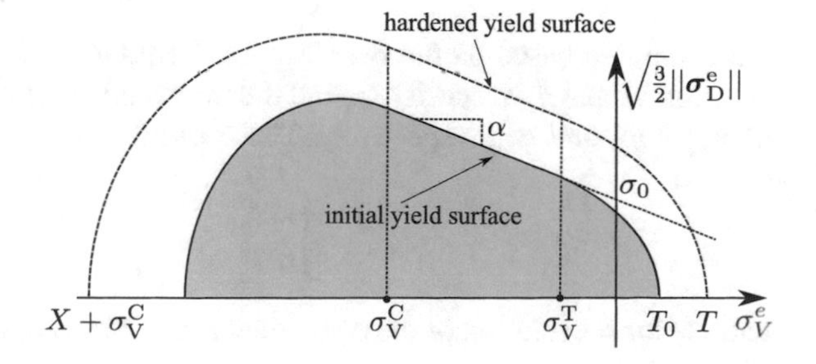

Fig. 10.4 Smooth three-surface microplane cap yield function

$$X = R f_{\mathrm{DP}}\left(\sigma_V^C\right) . \tag{10.10}$$

The abscissa of the intersection point between the tension cap and the DRUCKER-PRAGER yield function is denoted by σ_V^T, T_0 is the initial intersection of the cap with the volumetric axis, R_t governs the influence of the hardening on this point, the abscissa of the intersection point between the compression cap and the DRUCKER-PRAGER yield function is denoted by σ_V^C, R is the ratio between the volumetric and the deviatoric axes of the compression cap and $H(\bullet)$ denotes a HEAVISIDE step function. A linear hardening function

$$f_h(\kappa) = D\kappa \tag{10.11}$$

is applied, where D is a material constant and the hardening variable κ is related to the plastic multiplier λ by the evolution equation

$$\dot{\kappa} = \dot{\lambda} . \tag{10.12}$$

Damage on the microplane level

$$d^{\mathrm{mic}} = 1 - \left(1 - d_{\mathrm{c}}^{\mathrm{mic}}\right)\left(1 - r_{\mathrm{w}} d_{\mathrm{t}}^{\mathrm{mic}}\right) \tag{10.13}$$

is divided into tensile damage

$$d_{\mathrm{t}}^{\mathrm{mic}} = 1 - \exp\left(-\beta_{\mathrm{t}} \gamma_{\mathrm{t}}^{\mathrm{mic}}\right) \tag{10.14}$$

and compressive damage

$$d_{\mathrm{c}}^{\mathrm{mic}} = 1 - \exp\left(-\beta_{\mathrm{c}} \gamma_{\mathrm{c}}^{\mathrm{mic}}\right) , \tag{10.15}$$

considering the material constants β_{t} and β_{c}. Furthermore, the quantities

$$\gamma_{\mathrm{t}}^{\mathrm{mic}} = \begin{cases} \hat{\eta}_{\mathrm{t}}^{\mathrm{mic}} - \gamma_{\mathrm{t0}} & \hat{\eta}_{\mathrm{t}}^{\mathrm{mic}} > \gamma_{\mathrm{t0}} \\ 0 & \hat{\eta}_{\mathrm{t}}^{\mathrm{mic}} \leq \gamma_{\mathrm{t0}} \end{cases} \tag{10.16}$$

and

$$\gamma_{\mathrm{c}}^{\mathrm{mic}} = \begin{cases} \hat{\eta}_{\mathrm{c}}^{\mathrm{mic}} - \gamma_{\mathrm{c0}} & \hat{\eta}_{\mathrm{c}}^{\mathrm{mic}} > \gamma_{\mathrm{c0}} \\ 0 & \hat{\eta}_{\mathrm{c}}^{\mathrm{mic}} \leq \gamma_{\mathrm{c0}} \end{cases} \tag{10.17}$$

are computed based on the over-non-local equivalent strains for tension and compression, $\hat{\eta}_{\mathrm{t}}$ and $\hat{\eta}_{\mathrm{c}}$, respectively, and the threshold values for tensile and compressive damage, η_{t0} and η_{c0}, respectively. The evolution of the local equivalent strains for tension

$$\dot{\eta}_{\mathrm{t}}^{\mathrm{mic}} = \begin{cases} r_{\mathrm{w}} \dot{\varepsilon}_{\mathrm{V}}^{\mathrm{mic,pl}} & \dot{\varepsilon}_{\mathrm{V}}^{\mathrm{mic,pl}} > 0 \\ 0 & \dot{\varepsilon}_{\mathrm{V}}^{\mathrm{mic,pl}} \leq 0 \end{cases} \tag{10.18}$$

and the local evolution of the equivalent strain for compression

$$\dot{\eta}_{\mathrm{c}}^{\mathrm{mic}} = \begin{cases} (1 - r_{\mathrm{w}})\, \dot{\varepsilon}_{\mathrm{V}}^{\mathrm{mic,pl}} & \dot{\varepsilon}_{\mathrm{V}}^{\mathrm{mic,pl}} > 0 \\ 0 & \dot{\varepsilon}_{\mathrm{V}}^{\mathrm{mic,pl}} \leq 0 \end{cases} \tag{10.19}$$

at each microplane are directly related to the plastic parts of the local strains at the microplanes. Here, the split weight factor

$$r_{\mathrm{w}} = \frac{\sum_{I=1}^{3} \langle \varepsilon^I \rangle}{\sum_{I=1}^{3} |\varepsilon^I|} \tag{10.20}$$

evaluates the Ith principal value of the strain tensor ε^I and its positive part $\langle \varepsilon^I \rangle$. The non-local counterparts of the two components of the equivalent strain are governed by the differential equations

$$\bar{\eta}_{\mathrm{mt}} - c\nabla^2 \bar{\eta}_{\mathrm{mt}} = \frac{1}{4\pi} \int_{\Omega} \eta_{\mathrm{t}}^{\mathrm{mic}} \, d\Omega \tag{10.21}$$

and

$$\bar{\eta}_{\mathrm{mc}} - c\nabla^2 \bar{\eta}_{\mathrm{mc}} = \frac{1}{4\pi} \int_{\Omega} \eta_{\mathrm{c}}^{\mathrm{mic}} \, d\Omega \tag{10.22}$$

on the macro-level with the gradient activity parameter c. A linear combination of the local and the non-local equivalent strains based on a material parameter m yields the over-non-local components

$$\hat{\eta}_{\mathrm{mt}}^{\mathrm{mic}} = m\bar{\eta}_{\mathrm{mt}} + (1 - m)\, \eta_{\mathrm{t}}^{\mathrm{mic}} \tag{10.23}$$

and

$$\hat{\eta}_{\mathrm{mc}}^{\mathrm{mic}} = m\bar{\eta}_{\mathrm{mc}} + (1 - m)\, \eta_{\mathrm{c}}^{\mathrm{mic}}. \tag{10.24}$$

The return mapping algorithm is based on the trial stresses

$$\sigma_V^{\mathrm{mic,tr}} = K^{\mathrm{mic}} \left(\varepsilon_V^{\mathrm{mic}} - \varepsilon_V^{\mathrm{mic,pl,n\text{-}1}} \right) \tag{10.25}$$

and

$$\boldsymbol{\sigma}_D^{\mathrm{mic,tr}} = 2G^{\mathrm{mic}} \left(\boldsymbol{\varepsilon}_{\mathrm{D}}^{\mathrm{mic}} - \boldsymbol{\varepsilon}_{\mathrm{D}}^{\mathrm{mic,pl,n\text{-}1}} \right) \tag{10.26}$$

on the microplane level with respect to the plastic strains at the previous step $n - 1$. In case of a violation of the yield criterion given by Eq. (10.5) by the trial state, the system of equations

$$\begin{aligned} 0 &= \frac{3}{2} \frac{\left\| \boldsymbol{\sigma}_{\mathrm{D}}^{\mathrm{mic,tr}} \right\|^2}{(1 + 6\Delta\lambda^n G^{\mathrm{mic}})^2} - f_{\mathrm{DP}}^2 f_{\mathrm{c}} f_{\mathrm{t}} \\ 0 &= \sigma_{\mathrm{V}}^{\mathrm{mic}} - \sigma_{\mathrm{V}}^{\mathrm{mic,tr}} + \Delta\lambda^n K^{\mathrm{mic}} m_{\mathrm{V}} \end{aligned} \tag{10.27}$$

with the flow direction

$$m_{\mathrm{V}} = -2 f_{\mathrm{DP}} \frac{\partial f_{\mathrm{DP}}}{\partial \sigma_{\mathrm{V}}^{\mathrm{mic}}} f_{\mathrm{c}} f_{\mathrm{t}} - f_{\mathrm{DP}}^2 \frac{\partial f_{\mathrm{c}}}{\partial \sigma_{\mathrm{V}}^{\mathrm{mic}}} f_{\mathrm{t}} - f_{\mathrm{DP}}^2 f_{\mathrm{c}} \frac{\partial f_{\mathrm{t}}}{\partial \sigma_{\mathrm{V}}^{\mathrm{mic}}} \tag{10.28}$$

is solved for the increment of the plastic multiplier $\Delta\lambda^n$ and the volumetric stress $\sigma_{\mathrm{V}}^{\mathrm{mic}}$ by an iterative NEWTON method. The updated deviatoric stress, deviatoric plastic strain and volumetric plastic strain read

$$\boldsymbol{\sigma}_{\mathrm{D}}^{\mathrm{mic}} = \boldsymbol{\sigma}_{\mathrm{D}}^{\mathrm{mic,tr}} / \left(1 + 6\Delta\lambda^n G^{\mathrm{mic}}\right) , \tag{10.29}$$

$$\boldsymbol{\varepsilon}_{\mathrm{D}}^{\mathrm{mic,pl}} = \boldsymbol{\varepsilon}_{\mathrm{D}}^{\mathrm{mic}} - \frac{\boldsymbol{\sigma}_{\mathrm{D}}^{\mathrm{mic}}}{2G^{\mathrm{mic}}} \tag{10.30}$$

and

$$\varepsilon_{\mathrm{V}}^{\mathrm{mic,pl}} = \varepsilon_{\mathrm{V}}^{\mathrm{mic}} - \frac{\sigma_{\mathrm{V}}^{\mathrm{mic}}}{K^{\mathrm{mic}}} , \tag{10.31}$$

respectively.

The gradient enhanced microplane model with damage and plasticity outlined above includes all features necessary to model the characteristic behavior of SHCC at tensile loading. The yield function involves an initial yield limit σ_0 to approximate the onset of the first micro crack at a specific level of stress. The material parameter D, used to describe the linear hardening, can be calibrated to model the evolution of multiple micro cracks at large deformation. Finally, the damage approach is suitable to approximate softening behavior due to pull out of fibers observed in the experiments. Additionally, the model incorporates a number of beneficial aspects. Due to gradient enhancement, the well-known mesh sensitivity and numerical instability of softening damage models is overcome and localization of strains is prevented. The compression and tension caps in the yield function may not be vital for the modeling of SHCC at tension. However, they ensure a numerically stable return mapping algorithm in every case.

10.2.2 Phase-Field Model for Ductile Fracture

The phase-field approach for crack approximation is an emerging technique, due to two key aspects, that are beneficial for the numerical simulation of fracture within the framework of finite elements. The first aspect involves a continuous representation of a discrete crack topology by means of an additional field parameter – the phase-field p. In its representation as an order parameter, it is used to distinguish between sound ($p = 0$) and broken ($p = 1$) regions of a solid body. Specific spatial information about the crack topology Γ (orientation of discrete surfaces, position of the crack tip) are not available, see Fig. 10.5. Instead, the amount of regularized crack surface Γ_l can be calculated by the volume integration of the crack surface density

$$\gamma = \frac{1}{2l}(p^2 + l^2 \cdot |\nabla p|^2). \tag{10.32}$$

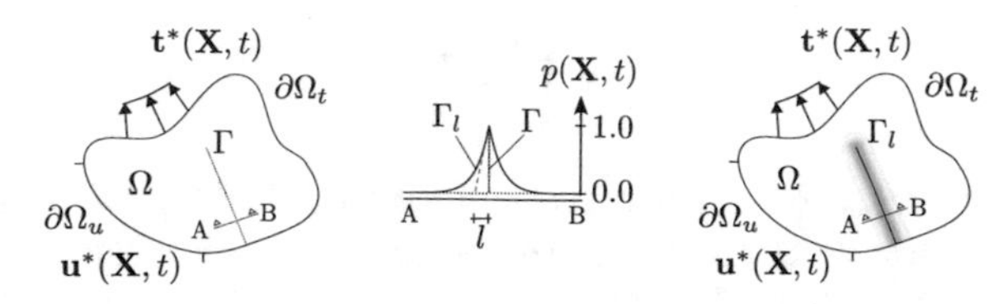

Fig. 10.5 Continuum Ω with sharp crack Γ and its continuous approximation Γ_l

Here, the width of the transition zone between sound and broken is governed by the length scale parameter l. A discrete crack topology may be recovered by

$$\Gamma = \lim_{l \to 0} \Gamma_l. \tag{10.33}$$

The second aspect is the degradation of a specific part of the strain energy density by means of a degradation function $g(p)$. The strain energy density with respect to the degradation reads

$$\psi_\varepsilon = g(p) \cdot \psi^+ + \psi^- , \tag{10.34}$$

where ψ^+ and ψ^- are the crack driving and the persistent components of the strain energy density, respectively. The energetic combination of both aspects results in a numerical approach for crack evolution, that obtains a number of important fracture phenomena (crack initiation in the bulk, crack propagation speed, branching phenomenon) in a straight forward manner, see e.g. Steinke et al (2016).

Plastic material behavior is considered by the plastic energy density

$$\psi_\mathrm{p} = \sigma_0 \kappa + \frac{1}{2} D \kappa^2. \tag{10.35}$$

One approach to model the transition from brittle to ductile fracture is published in Ambati et al (2015). A modified degradation function

$$g(p, q) = (1 - p)^{2q} \tag{10.36}$$

is introduced, where $q = \varepsilon_\mathrm{eq}^\mathrm{pl} / \varepsilon_\mathrm{eq}^\mathrm{crit}$ is an additional multiplier to the exponent of the degradation function, that is related to the evolution of plastic strains

$$\varepsilon_\mathrm{eq}^\mathrm{pl} = \int_0^t ||\dot{\boldsymbol{\varepsilon}}^\mathrm{pl}|| \,\mathrm{d}\tau \tag{10.37}$$

and a threshold value $\varepsilon_\mathrm{eq}^\mathrm{crit}$. A von Mises yield function

$$f_\mathrm{VM} = ||\boldsymbol{\sigma}_\mathrm{D}|| - t_\kappa \tag{10.38}$$

with the hardening thermodynamical force

$$t_\kappa = \sigma_0 + f_h(\kappa) \tag{10.39}$$

is considered. The definition of the crack driving strain energy density

$$\psi^+_{\text{VD}} = \left(\frac{\lambda}{2} + \frac{\mu}{3}\right) \langle \mathbf{V} : \boldsymbol{\varepsilon} \rangle^2 + \mu\,(\mathbf{Dev} : \boldsymbol{\varepsilon}) : (\mathbf{Dev} : \boldsymbol{\varepsilon}) \tag{10.40}$$

is based on the volumetric-deviatoric decomposition of the strains introduced in Amor et al (2009). Here, the strain energy related to volumetric expansion and deviatoric strains is assumed to drive the crack's evolution. In this case, an analytic solution for the updated hardening variable and the plastic strains is possible. An alternative decomposition

$$\psi^+_{\text{S}} = \frac{\lambda}{2} \langle \mathbf{V} : \boldsymbol{\varepsilon} \rangle^2 + \mu \mathbf{V} : (\boldsymbol{\varepsilon}_+ \cdot \boldsymbol{\varepsilon}_+), \tag{10.41}$$

proposed in Hofacker et al (2009), is based on the Ith eigenvalues ε^I and eigenvectors $\mathbf{n_I}$ of the strain tensor, where $\boldsymbol{\varepsilon}_+ = \sum_{I=1}^{3} \langle \varepsilon^I \rangle\ \mathbf{n_I} \otimes \mathbf{n_I}$. Recently, a directional split of the strain energy density has been proposed in Steinke and Kaliske (2018). Essentially, in contrast to the volumetric-deviatoric and the spectral decompositions, a realistic post-fracture behavior of the phase-field crack is obtained by the directional approach. However, analytical solutions for the updated hardening variable and the plastic strains are not available any more. Instead, a return mapping algorithm based on an iterative Newton method is applied. In the following, the basic aspects of the directional split are recapitulated and classified within the framework of ductile fracture with a phase-field model. Then, the return mapping algorithm is specified.

The regularized representation of the crack topology by the phase-field provides essential benefits for the simulation of fracture within the framework of the finite element method. However, as a consequence, the spatial orientation of the crack is not available. Therefore, basic kinematic characteristics of a crack (transmission of compressive forces via contact, no tensile stresses or shear stresses on the crack surface) cannot be recovered properly. The directional phase-field approach is based on the decomposition of the ground stress tensor

$$\boldsymbol{\sigma}_0 = \lambda \cdot \left[\left(\boldsymbol{\varepsilon} - \boldsymbol{\varepsilon}^{\text{pl}}\right) : \mathbf{1}\right] \mathbf{1} + 2\mu \left(\boldsymbol{\varepsilon} - \boldsymbol{\varepsilon}^{\text{pl}}\right) \tag{10.42}$$

with respect to a crack reference coordinate system (CCS). The CCS is an orthonormal coordinate system with the axes **r**, **s** and **t**, where **r** is perpendicular on the crack surface, see Fig. 10.6. A set of projection tensors

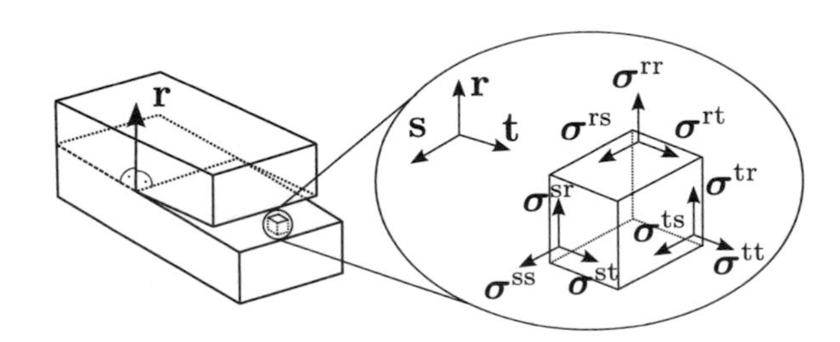

Fig. 10.6 Crack orientation vector **r** and decomposition of a stress tensor in the orthonormal crack reference coordinate system (CCS)

$$\begin{aligned} &\boldsymbol{M}^{\mathrm{rr}} = \mathbf{r} \otimes \mathbf{r}, && \boldsymbol{M}^{\mathrm{rs}} = \mathbf{r} \otimes \mathbf{s}, && && \boldsymbol{M}^{\mathrm{rt}} = \mathbf{r} \otimes \mathbf{t}, \\ &\boldsymbol{M}^{\mathrm{sr}} = \mathbf{s} \otimes \mathbf{r}, && \boldsymbol{M}^{\mathrm{ss}} = \mathbf{s} \otimes \mathbf{s}, && && \boldsymbol{M}^{\mathrm{st}} = \mathbf{s} \otimes \mathbf{t}, \\ &\boldsymbol{M}^{\mathrm{tr}} = \mathbf{t} \otimes \mathbf{r}, && \boldsymbol{M}^{\mathrm{ts}} = \mathbf{t} \otimes \mathbf{s} && \text{and} && \boldsymbol{M}^{\mathrm{tt}} = \mathbf{t} \otimes \mathbf{t} \end{aligned} \tag{10.43}$$

is defined in order to obtain the single components

$$\begin{aligned} &\boldsymbol{\sigma}^{\mathrm{rr}} = \hat{\sigma}^{\mathrm{rr}} \boldsymbol{M}^{\mathrm{rr}}, && \boldsymbol{\sigma}^{\mathrm{rs}} = \hat{\sigma}^{\mathrm{rs}} \boldsymbol{M}^{\mathrm{rs}}, && && \boldsymbol{\sigma}^{\mathrm{rt}} = \hat{\sigma}^{\mathrm{rt}} \boldsymbol{M}^{\mathrm{rt}}, \\ &\boldsymbol{\sigma}^{\mathrm{sr}} = \hat{\sigma}^{\mathrm{sr}} \boldsymbol{M}^{\mathrm{sr}}, && \boldsymbol{\sigma}^{\mathrm{ss}} = \hat{\sigma}^{\mathrm{ss}} \boldsymbol{M}^{\mathrm{ss}}, && && \boldsymbol{\sigma}^{\mathrm{st}} = \hat{\sigma}^{\mathrm{st}} \boldsymbol{M}^{\mathrm{st}}, \\ &\boldsymbol{\sigma}^{\mathrm{tr}} = \hat{\sigma}^{\mathrm{tr}} \boldsymbol{M}^{\mathrm{tr}}, && \boldsymbol{\sigma}^{\mathrm{ts}} = \hat{\sigma}^{\mathrm{ts}} \boldsymbol{M}^{\mathrm{ts}} && \text{and} && \boldsymbol{\sigma}^{\mathrm{tt}} = \hat{\sigma}^{\mathrm{tt}} \boldsymbol{M}^{\mathrm{tt}} \end{aligned} \tag{10.44}$$

based on the scalar magnitude of each stress component

$$\begin{aligned} &\hat{\sigma}^{\mathrm{rr}} = \boldsymbol{\sigma}_0 : \boldsymbol{M}^{\mathrm{rr}}, && \hat{\sigma}^{\mathrm{rs}} = \boldsymbol{\sigma}_0 : \boldsymbol{M}^{\mathrm{rs}}, && && \hat{\sigma}^{\mathrm{rt}} = \boldsymbol{\sigma}_0 : \boldsymbol{M}^{\mathrm{rt}}, \\ &\hat{\sigma}^{\mathrm{sr}} = \boldsymbol{\sigma}_0 : \boldsymbol{M}^{\mathrm{sr}}, && \hat{\sigma}^{\mathrm{ss}} = \boldsymbol{\sigma}_0 : \boldsymbol{M}^{\mathrm{ss}}, && && \hat{\sigma}^{\mathrm{st}} = \boldsymbol{\sigma}_0 : \boldsymbol{M}^{\mathrm{st}}, \\ &\hat{\sigma}^{\mathrm{tr}} = \boldsymbol{\sigma}_0 : \boldsymbol{M}^{\mathrm{tr}}, && \hat{\sigma}^{\mathrm{ts}} = \boldsymbol{\sigma}_0 : \boldsymbol{M}^{\mathrm{ts}} && \text{and} && \hat{\sigma}^{\mathrm{tt}} = \boldsymbol{\sigma}_0 : \boldsymbol{M}^{\mathrm{tt}}. \end{aligned} \tag{10.45}$$

The total stress

$$\boldsymbol{\sigma} = g(p,q) \cdot \boldsymbol{\sigma}^{+} + \boldsymbol{\sigma}^{-} \tag{10.46}$$

is obtained by the degradation of the dissolved stress components $\boldsymbol{\sigma}^{+}$, that are affected by the crack, and the persistent part $\boldsymbol{\sigma}^{-}$, that is undegraded. The dissolved stress components

$$\begin{aligned} \boldsymbol{\sigma}^{+} = {} & \langle \hat{\sigma}^{\mathrm{rr}} \rangle \boldsymbol{M}^{\mathrm{rr}} + \boldsymbol{\sigma}^{\mathrm{rs}} + \boldsymbol{\sigma}^{\mathrm{sr}} + \boldsymbol{\sigma}^{\mathrm{rt}} + \boldsymbol{\sigma}^{\mathrm{tr}} \\ & + \frac{\lambda}{\lambda + 2\mu} \langle \hat{\sigma}^{\mathrm{rr}} \rangle \left(\boldsymbol{M}^{\mathrm{ss}} + \boldsymbol{M}^{\mathrm{tt}} \right) \end{aligned} \tag{10.47}$$

are directly related to the basic kinematic characteristics of a crack, i.e. tension and shear on the crack surface are considered. Additionally, a term with respect to Poisson's effect is required. The persistent part reads

$$\begin{aligned} \boldsymbol{\sigma}^{-} = {} & \frac{\hat{\sigma}^{\mathrm{rr}} - |\hat{\sigma}^{\mathrm{rr}}|}{2} \boldsymbol{M}^{\mathrm{rr}} + \boldsymbol{\sigma}^{\mathrm{ss}} + \boldsymbol{\sigma}^{\mathrm{tt}} + \boldsymbol{\sigma}^{\mathrm{st}} + \boldsymbol{\sigma}^{\mathrm{ts}} \\ & - \frac{\lambda}{\lambda + 2\mu} \langle \hat{\sigma}^{\mathrm{rr}} \rangle \left(\boldsymbol{M}^{\mathrm{ss}} + \boldsymbol{M}^{\mathrm{tt}} \right). \end{aligned} \tag{10.48}$$

The time discretizations of the evolution equation for the plastic strains

$$\dot{\boldsymbol{\varepsilon}}^{\mathrm{pl}} = \dot{\lambda} \frac{\partial f_{\mathrm{VM}}}{\partial \boldsymbol{\sigma}_{\mathrm{D}}} \quad \Rightarrow \quad \mathbf{R}_1 = 0 = \boldsymbol{\varepsilon}^{\mathrm{pl,n}} - \boldsymbol{\varepsilon}^{\mathrm{pl,n\text{-}1}} - \Delta\lambda \cdot \frac{\partial f_{\mathrm{VM}}}{\partial \boldsymbol{\sigma}_{\mathrm{D}}} \tag{10.49}$$

and the evolution equation for the hardening variable

$$\dot{\kappa} = \dot{\lambda} \frac{\partial f_{\mathrm{VM}}}{\partial t_{\kappa}} \quad \Rightarrow \quad R_2 = 0 = \kappa^{\mathrm{n}} - \kappa^{\mathrm{n\text{-}1}} - \Delta\lambda \cdot \frac{\partial f_{\mathrm{VM}}}{\partial t_{\kappa}} \tag{10.50}$$

as well as the evaluation of the yield function Eq. (10.38) based on the deviatoric component of the stress tensor

$$\boldsymbol{\sigma}_{\mathrm{D}} = \mathbf{Dev} : \boldsymbol{\sigma} \tag{10.51}$$

are considered in order to establish the nonlinear equation system

$$0 = \begin{bmatrix} \mathbf{R}_1 \\ R_2 \\ f_{\mathrm{VM}} \end{bmatrix} + \begin{bmatrix} \frac{\partial \mathbf{R}_1}{\partial \boldsymbol{\varepsilon}^{\mathrm{p}}} & \frac{\partial \mathbf{R}_1}{\partial \kappa} & \frac{\partial \mathbf{R}_1}{\partial \Delta\kappa} \\ \frac{\partial R_2}{\partial \boldsymbol{\varepsilon}^{\mathrm{pl}}} & \frac{\partial R_2}{\partial \kappa} & \frac{\partial R_2}{\partial \Delta\kappa} \\ \frac{\partial f_{\mathrm{VM}}}{\partial \boldsymbol{\varepsilon}^{\mathrm{pl}}} & \frac{\partial f_{\mathrm{VM}}}{\partial \kappa} & \frac{\partial f_{\mathrm{VM}}}{\partial \Delta\kappa} \end{bmatrix} \begin{bmatrix} \Delta\boldsymbol{\varepsilon}^{\mathrm{pl}} \\ \Delta\kappa \\ \Delta\Delta\lambda \end{bmatrix}, \tag{10.52}$$

that is solved for the increments for the plastic strain, the hardening variable and the increment of the plastic multiplier, $\Delta\boldsymbol{\varepsilon}^{\mathrm{pl}}$, $\Delta\kappa$ and $\Delta\Delta\lambda$ by an iterative NEWTON scheme.

The framework for a phase-field model for ductile fracture outlined above includes all features necessary to model the characteristic behavior of SHCC under tensile loading. In analogy to the microplane approach, the initial yield limit σ_0 can be used to specify the stress level for the onset of the first micro crack and the hardening modulus D is used to model the behavior during the evolution of multiple micro cracks. In contrast to the microplane model, the phase-field approach approximates the evolution of cracks. Hence, a more abrupt drop in the reaction forces is obtained, which corresponds to rupture of fibers bridging a micro crack. A very simple VON MISES yield function without caps for tension and compression is applied in order to show a proof of concept for this new modeling technique. A major difference is the fact, that the evaluation of the yield criterion is based on the degraded stresses for the phase-field and the microplane model considers undamaged stress components. Furthermore, the numerical representation of a crack by the phase-field approach requires a finite element discretization with very small elements.

10.3 Dumbbell Tension Test

A detailed investigation of the tensile behavior of SHCC based on experimental data for low, intermediate and high strain rates is given in Curosu (2017). For the low strain rate experiments, a dumbbell specimen is considered, see Fig. 10.2a for the geometry of the specimen. An Instron 8501 testing machine has been used and the load has been applied in deformation controlled mode. The typical measurements of the relation between strains and stresses, see e.g. Figs. 10.3b and 10.3c, are based on a force sensor, that has been fixed to the upper stationary cross head, and the data from two linear variable differential transformers (LVDT) in a steel frame attached to the specimen, see Fig. 10.2a. The specimen has been glued to the testing machine in order to apply the tensile loading. The deformation originates from a hydraulic actuator. A set of 4 different deformation velocities, i.e. 0.05 mm/s, 0.5 mm/s,

1 mm/s and 20 mm/s, has been tested. Therefore, the mean strain rate over the total height of the specimen is in between 0.1/s and 0.2/ms.

The evaluation of the experiments yields results for the tensile strength, the ultimate strain, the amount of work-to-fracture and diagrams of the relation between stresses and strains on the structural level. Stresses are computed by dividing the reaction forces of the force sensor by the area of the cross-section of the specimen glued to the machine ($A = 40\ \text{mm} \cdot 40\ \text{mm} = 16\ \text{cm}^2$). Strains are actually mean strains, that are computed by the average of the deformations measured by the two LVDTs divided by the total height of the specimen. The tensile strength denotes the highest stress value, that is reached before the onset of the first micro crack. The ultimate strain indicates the end of the micro crack evolution, i.e. the point when fiber pull out or rupture takes place. The amount of work-to-fracture is the energy, that is introduced into the system up to the point where the ultimate strain is reached. It is equal to the area below the graph in the diagrams for the relation between stresses and strains excluding the softening region. A subset of the diagrams for the relation between stresses and strains published in Curosu (2017) are given for fiber reinforcement with PVA (polyvinyl-alcohol) and PE (polyethylene) in Figs. 10.3b and 10.3c, respectively. The different materials of the fibers have strong influence on the chemical bond between the cement matrix and the fibers. This results in differences, both for the evolution of the micro cracks as well as the structural failure. The characteristic behavior represented by the set linear-elasticity – hardening – softening/rupture, as shown in Fig. 10.2b, can be observed for both diagrams. Here, PVA exhibits a softening-like characteristic for the structural failure, i.e. the fibers are subsequently pulled out of the matrix, which results in a decreased force transmitted over the structure. The reinforcement by PE shows an abrupt drop in the reaction forces, that indicates the sudden rupture of the fibers and a total loss of the structural integrity. Actually, this is true for a limited amount of specimen with PE reinforcement, as there is experimental evidence for a long range of softening due to fiber pull out for these specimen, too. However, from the experimental point of view, it is not clear yet how to explain and/or predict the failure type. Nevertheless, this contribution is focused on the demonstration, that both failure mechanisms can be approximated numerically by the choice of the proper model.

10.3.1 Damage-like Failure - Microplane Results

The simulations with the microplane model are focused on the approximation of the experiments, where the structural failure exhibits a damage-like characteristic with a relatively long period of softening due to the pull out of the fibers. Model parameters according to Table 10.1 are assumed.

The numerical setup contains a finite element discretization of the specimen in 3D with 1032 8-node brick elements with linear shape functions. Displacement boundaries are imposed on the upper and lower face of the specimen in all directions and monotonically increasing displacements in vertical direction up to a final

Table 10.1 Model parameter for the microplane approach

$E =$ 15 GPa	$\nu =$ 0.2	$D =$ 800 MPa^2
$\sigma_{\mathrm{V}}^{\mathrm{C}} = -40$ MPa	$R =$ 2	$c =$ 36 mm^2
$m =$ 2.5	$\beta_{\mathrm{c}} = 200$	$\beta_{\mathrm{t}} = 250$

displacement of $u = 5$ mm are applied. A quasi-static simulation is performed, i.e. transient effects are neglected. The results for different strain rates are obtained via calibration of the parameters for the uniaxial compressive strength f_c and for the threshold for tensile damage γ_t. It should be noted, that both the friction angle α and the initial yield stress σ_0 are directly related to the uniaxial compressive strength by

$$\alpha = 0.2 \qquad \text{and} \qquad \sigma_0 = \frac{\sqrt{3} - \alpha}{3} f_c,$$

see e.g. Zreid and Kaliske (2018) for a detailed explanation of these relations in the framework of concrete. The result of the calibration of the two parameters is summarized in Table 10.2. Given these parameters, a good agreement to the experimentally observed relations between stresses and strains can be obtained, see Fig. 10.7. The initial elastic response is approximated exactly. In the experiments, the onset of the first micro crack is accompanied with a significant drop of the stresses. Furthermore, the subsequent evolution of multiple cracks also shows (smaller) decreases of the stresses and a tooth saw pattern is observed. As this mechanism is not part of the model, this characteristic is not reproduced exactly. Instead, the linear hardening yields a good approximation of the global response. Finally, the numerical simulation of the softening shows a very good agreement to the experimental data.

According to the concept of a dynamic increase factor (DIF), the strain rate dependency of the model parameters is described by functions with respect to the strain rate, see Fig. 10.8. Based on the simulation results, the strain rate dependent value of the threshold for tensile damage is given by

$$\gamma_t(\dot{\varepsilon}) = 0.05 \cdot \ln(\dot{\varepsilon} \cdot \mathrm{s}) + 3.48 \tag{10.53}$$

and the strain rate dependent value of the uniaxial compressive strength is given by

Table 10.2 Calibration of the strength and the threshold for tensile damage with respect to the strain rate

$\dot{\varepsilon}$ in 1/ms	f_c in MPa	γ_t in %
100	90	3.10
4.0	60	3.00
2.0	50	2.80
0.2	40	2.75

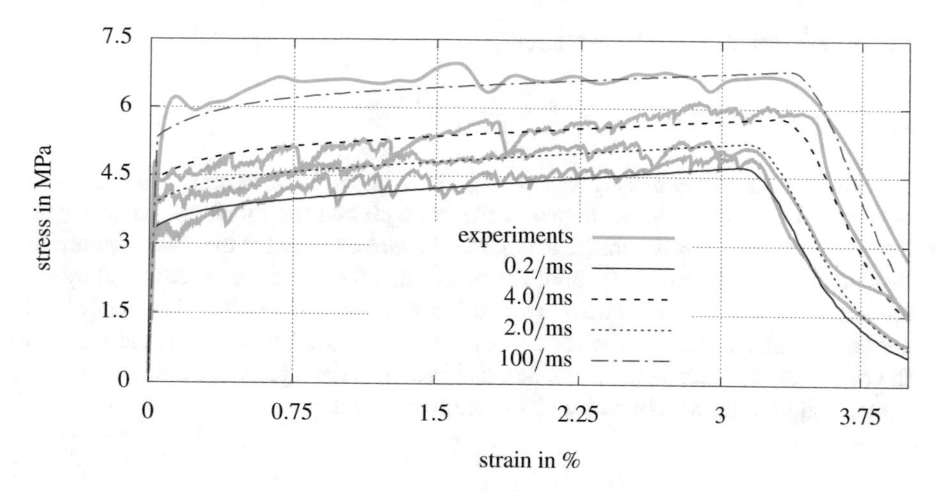

Fig. 10.7 Comparison of the experimental data and the simulation results for the relation between stresses and strains for different rates of strain

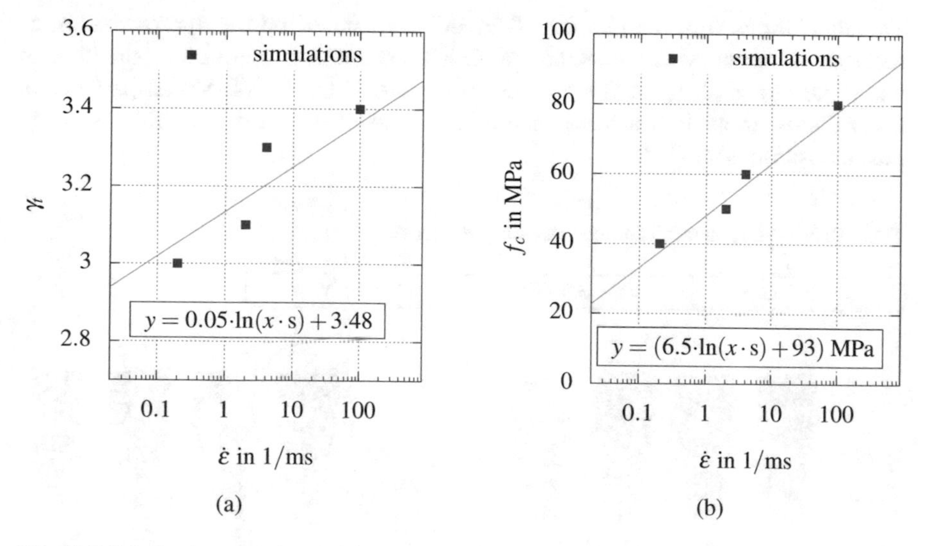

Fig. 10.8 Relation between the model parameters ((a) - threshold for tensile damage γ_t and (b) - uniaxial compressive strength f_c) and the strain rate

$$f_c(\dot{\varepsilon}) = 6.5 \text{ MPa} \cdot \ln(\dot{\varepsilon} \cdot \text{s}) + 93 \text{ MPa}. \tag{10.54}$$

In analogy to standard assumptions for the DIF, increasing strain rates yield increasing values for the strength (or similar values).

Contour plots for the homogenized damage

$$d^{\text{hom}} = \frac{1}{4\pi} \int_{\Omega} d^{\text{mic}} d\Omega$$

and the homogenized hardening variable

$$\kappa^{\text{hom}} = \frac{1}{4\pi}\int_{\Omega} \kappa^{\text{mic}} d\Omega$$

for different strain rates are given in Fig. 10.9 at the final vertical displacement of 5 mm. Due to the higher values of the strength and the threshold value for an increased strain rate, less damage and hardening are observed at the same amount of displacement. Furthermore, it should be noted, that the plasticity, represented by the hardening variable, is distributed in a relatively uniform way before the onset of the softening. Then, as soon as the softening takes place, localization of the plasticity in the center of the damage zone can be observed. The damage is concentrated in the center region of the specimen for all different strain rates.

10.3.2 Fracture-like Failure - Phase-Field Results

The simulations with the phase-field model are focused on the approximation of those experiments, where the structural failure exhibits a fracture-like characteristic with a short and abrupt softening due to rupture of the fibers. Model parameters according to Table 10.3 are assumed and the model with a volumetric-deviatoric decomposition is applied.

Table 10.3 Model parameter for the phase-field approach

λ = 8.9 GPa	μ = 13.3 GPa	D = 10 MPa

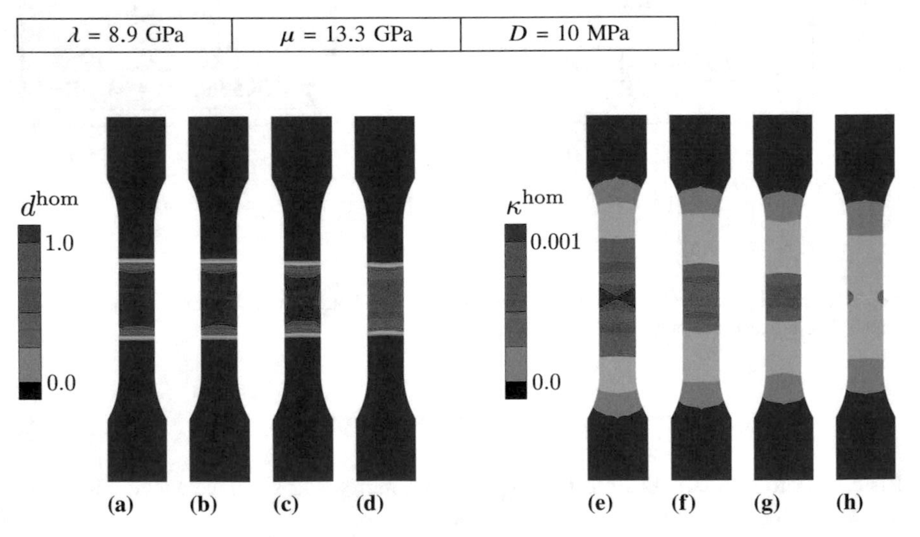

Fig. 10.9 Contour plots of the homogenized values for the damage and the hardening variable, d^{hom} and κ^{hom}, respectively, at a final displacement of 5 mm for strain rates 0.2/ms ((a), (e)), 2/ms ((b), (f)), 4/ms ((c), (g)) and 100/ms ((d), (h))

In analogy to the previous microplane simulations, phase-field model parameters are calibrated in order to fit the experimental data. Then, DIF are derived from the results of the quasi-static simulations. In the phase-field model for ductile fracture, the initial yield strength σ_0 is specified directly and affects the stress, where the structural response deviates from linear elasticity to plastic hardening. The numerical approximation of the structural failure is governed by the evolution of the phase-field, that is affected by two parameters, i.e. the fracture toughness G_c and the threshold value of the critical equivalent plastic strain $\varepsilon_{\text{eq}}^{\text{crit}}$. Increasing either of these two parameters results in a delayed evolution of the phase-field, i.e. the phase-field evolves at larger strains. Therefore, two sets of parameter calibration are performed. For the first set of calibration, a constant critical equivalent plastic strain is assumed and the fracture toughness is calibrated, see Fig. 10.10 for the results. In the second set of calibration, a constant fracture toughness is assumed and the critical equivalent plastic strain is calibrated, see Fig. 10.11 for the results.

As already observed for the microplane approach, the initial linear elasticity can be approximated exactly. A major difference between both approaches is observed in the behavior directly after plastic hardening starts to govern the response. While the microplane approach shows a relatively long period of decreasing hardening modulus for the structural response, the phase-field results exhibit a sharp transition between linear elasticity and plastic hardening. At this point, fitting the experimental data with the phase-field model is a challenging task, because a gradual decay of the hardening modulus, as observed for the microplane, is a very good approximation of the actual global response observed in the experiments, see Fig. 10.12. For the phase-field with the strictly bi-linear response, two approaches are available. The first option is to calibrate the initial yield stress such, that the deviation from the linear response in numerics coincides with the evolution of the first micro crack in

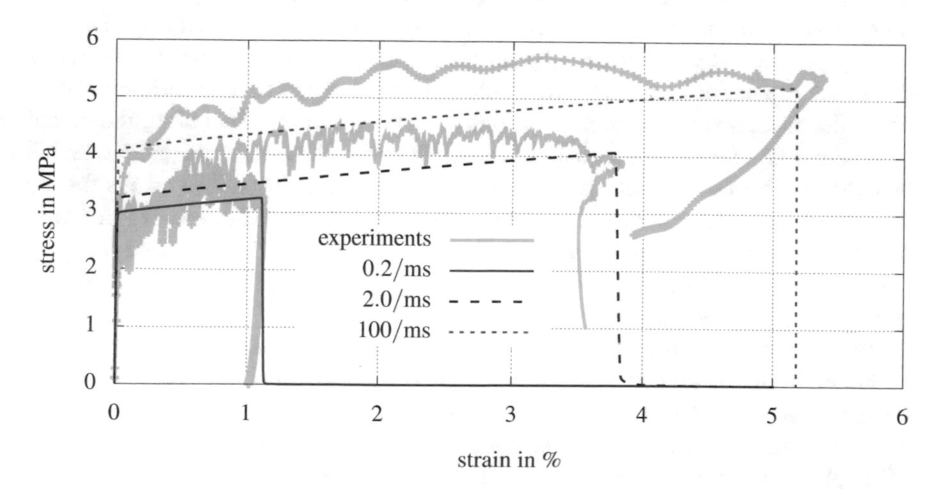

Fig. 10.10 Comparison of the experimental data and the simulation results for the relation between stresses and strains for different rates of strain at a constant threshold value of the critical equivalent plastic strain $\varepsilon_{\text{eq}}^{\text{crit}} = 4\ \%$

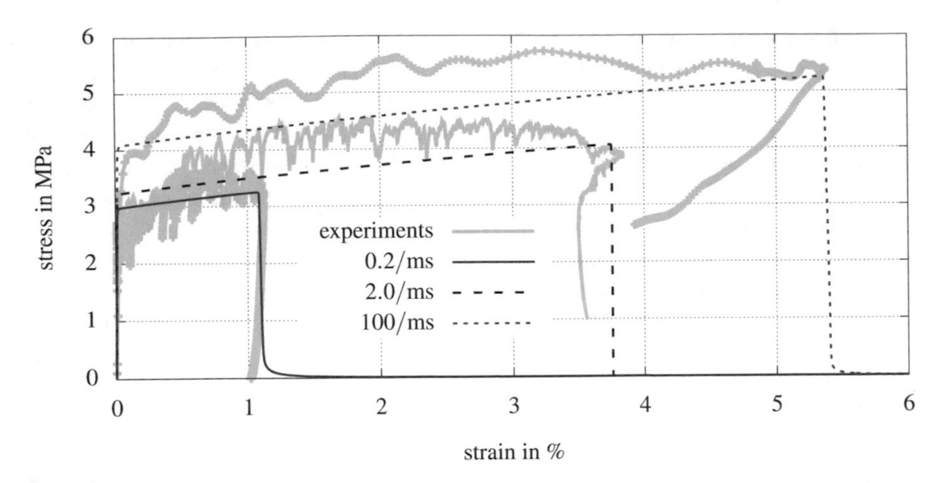

Fig. 10.11 Comparison of the experimental data and the simulation results for the relation between stresses and strains for different rates of strain at a constant fracture toughness $G_c = 40 \text{ J/m}^2$

the experiment. Then, the global hardening modulus is underestimated at first and overestimated for later state of multiple micro crack evolution in order to meet the stress value at the ultimate strain. The second option is to calibrate a hardening modulus as close as possible to the late state of multiple micro crack evolution. However, this would have resulted in a significant overestimation of the peak stress, where the first micro crack is evolved. Therefore, the first option has been chosen. Nevertheless, a non-linear hardening approach may be required in order to enhance the model and meet both the peak stress for the onset of the first micro crack as well as the global characteristic of the subsequent evolution of multiple micro cracks represented by the hardening. Finally, the experimentally observed structural failure of the specimen is characterized by an abrupt drop in the load bearing behavior due to the rupture of fibers. However, detailed and representative experimental data on the localization and the process of failure is not available. Therefore, the model parameters are calibrated such, that the evolution of the phase-field coincides with the experimentally observed ultimate strains on the structural level. The results for the parameter calibration of the initial yield stress, the fracture toughness at a constant threshold value of the critical equivalent plastic strain and the threshold value of the critical equivalent plastic strain at a constant value of the fracture toughness are summarized in Table 10.4.

The numerical setup contains a parameterized finite element discretization of the specimen in 3D by 8-node bricks with linear shape functions based on the natural number n, see Fig. 10.13. For the phase-field, a fine discretization is important in order to resolve the crack topology. Furthermore, the quality of the approximation of the phase-field profile, i.e. the typical shape of the transition zone between broken ($p = 1$) and sound ($p = 0$), in finite elements with linear shape functions depends on the relation between the element size and the length scale parameter l. In all phase-field simulations the relation

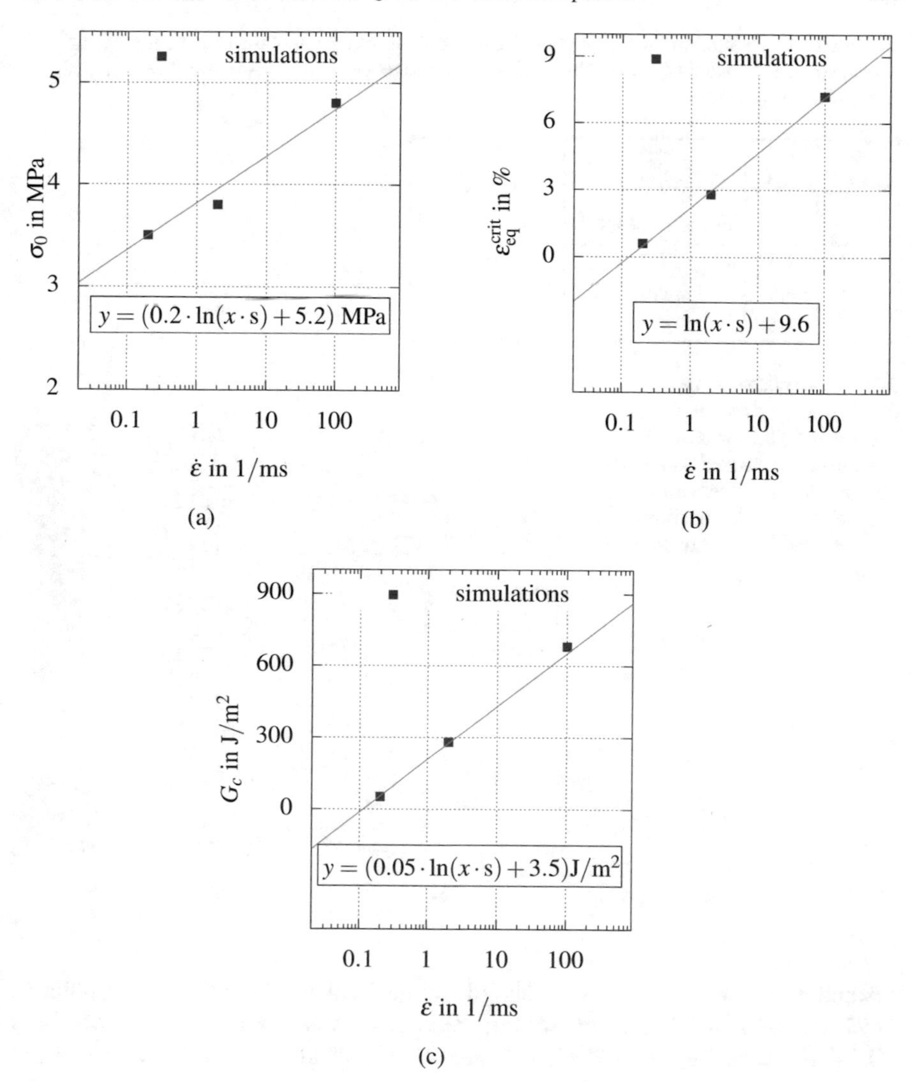

Fig. 10.12 Relation between the model parameters ((a) - initial yield stress σ_0, (b) - critical equivalent plastic strain ε_{eq}^{crit} and (c) - fracture toughness G_c) and the strain rate

$$l = \frac{12\ \text{mm}}{n} \tag{10.55}$$

is applied. In contrast to the microplane model, the phase-field approach aims at the approximation of cracks, i.e. the location of the crack initiation and the final topology can be obtained and evaluated. Furthermore, the phase-field approach for ductile fracture results in phase-field evolution in the vicinity of zones, that are highly deformed by plastic strains, i.e. the equivalent plastic strain is high. Based on the

Table 10.4 Calibration of the initial yield stress, the threshold value of the critical equivalent plastic strain at constant G_c = 40 J/m^2 and the fracture toughness at constant ε_{eq}^{crit} = 4 % with respect to the strain rate

$\dot{\varepsilon}$ in 1/ms	σ_0 in MPa	ε_{eq}^{crit} in % (at G_c = 40 J/m^2)	G_c in J/m^2 (at ε_{eq}^{crit} = 4 %)
100	4.8	7.2	650
2.0	3.8	2.8	280
0.2	3.5	0.6	52

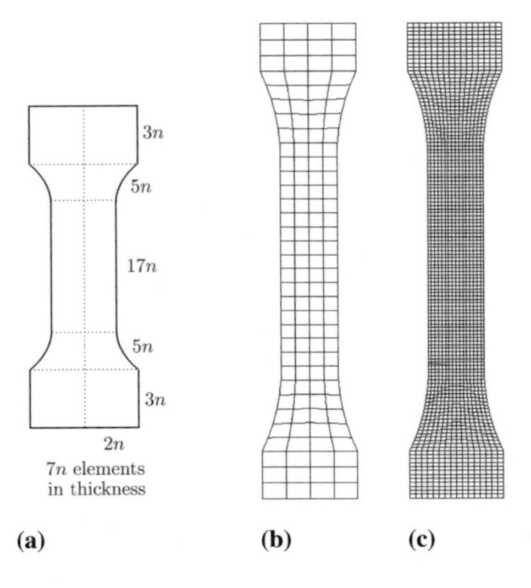

Fig. 10.13 Finite element discretization of the dumbbell specimen: (a) Schematic of the parameterized discretization concept based on the natural number n, (b) discretization for n = 1 and (c) discretization for n = 4

resolution of the discretization, a significant qualitative difference for the point of crack initiation can be observed, see Fig. 10.14. In a coarse mesh, see Figs. 10.14a and 10.14f, the hardening is relatively uniform over the slender part of the specimen and the phase-field evolves in a pattern similar to the results for the microplane approach, see Fig. 10.9 for comparison. In contrast, a higher resolution of the discretization (e.g. n = 4) results in two concentrations of the hardening variable, see Fig. 10.14g, and a subsequent crack initiation at these positions, see Figs. 10.14c, 10.14d and 10.14e.

10.4 Post-Fracture Crack Approximation with Phase-Field

The phase-field approach is a powerful tool for the realistic prediction of experimentally observed crack topologies in brittle, ductile and viscous fracture, see e.g. Ambati et al (2015); Loew et al (2019) and Steinke et al (2016). The reason for

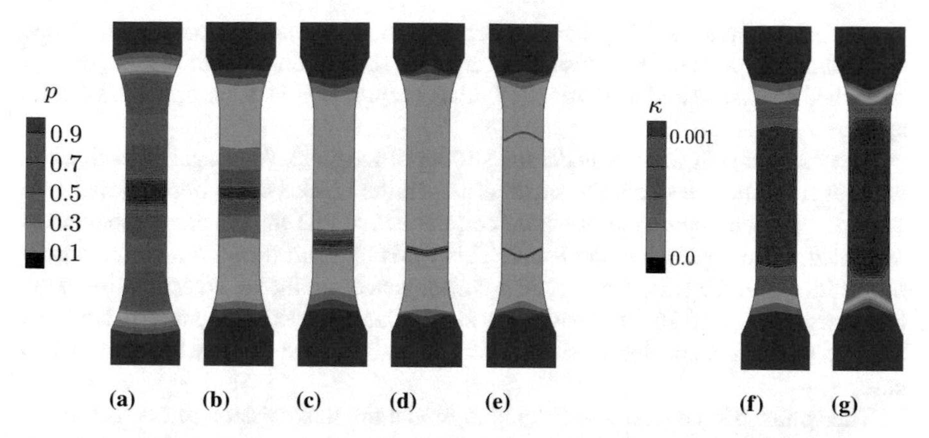

Fig. 10.14 Countour plots of the phase-field p at the end of crack evolution and the hardening variable κ before the onset of crack evolution for different finite element discretizations: (a) $n = 1$, (b) $n = 2$, (c) $n = 4$, (d) $n = 10$, (e) $n = 20$, (f) $n = 1$ and (g) $n = 10$

its outstanding performance is the basic assumption for crack evolution within the phase-field framework, that is motivated by fundamental physical principles. Generally, the formation of a crack surface is the result of the dissipation of strain energy. Here, a material-specific energetic level constitutes a barrier, that needs to be overcome in order to initiate or propagate a crack. Therefore, the concentration or spatial distribution of strain energy due to a specific geometry or loading situation automatically results in phenomena like branching, kinking and arrest, that are observed in a similar manner experimentally. One aspect, that is already considered in early phase-field models, is the fact, that the energetic level as a scalar quantity does not prevent the evolution of cracks due to the strain energy of a local, compressive strain. The first approach, to deal with this problem, is the so-called spectral decomposition of the strain energy density according to Eq. (10.41). The decomposition of the strain tensor into a sum of principal components enables the distinction between tensile and compressive components. A second approach involves the decomposition of the strain tensor into volumetric and deviatoric components, see Eq. (10.40). Here, strain energy due to volume reduction is not considered for the crack driving force.

As already stated, both approaches show very good results for the approximation of experimentally observed crack topologies at monotonic loading. However, within the framework of the variational principle, the split of the energy has a direct impact on the relation between stresses and strains. As a consequence, both approaches fail to reproduce a correct post-fracture behavior. The post-fracture behavior of a structure depends on the correct approximation of the kinematic characteristics of a crack. From a structural viewpoint, a crack is the separation of bulk material. Considering an ideal plane crack without friction, three basic characteristics are identified, see Fig. 10.15. On the one hand side, the crack faces can separate or slide against each other without resistance. On the other hand, a closed crack can transmit contact forces perpendicular to the crack faces. Both approaches, the spectral split as well as the volumetric-deviatoric decomposition, are not reproducing all three characteristics properly at the same time. Instead, the directional decomposition published in Steinke and Kaliske (2018) and recapitulated in Subsect. 10.2.2 can be applied.

The spectral split fails to model the sliding of the crack faces against each other without resistance. Instead, the shear strain on the crack face is decomposed into a tensile and a compressive principal component. While the tensile component is degraded, the compressive one is not. This leads to a significant amount of forces transmitted over the crack surface. As a consequence, additional cracks evolve at the boundary, see Fig. 10.16 for a benchmark simulation on an existing crack. In contrast, the directional split enables the simulation of a realistic post-fracture behavior in this case.

The volumetric-deviatoric split fails to model the transmission of contact forces perpendicular to the crack surfaces. The problem is obvious for the decomposition of an uniaxial compressive strain into its volumetric and deviatoric components by

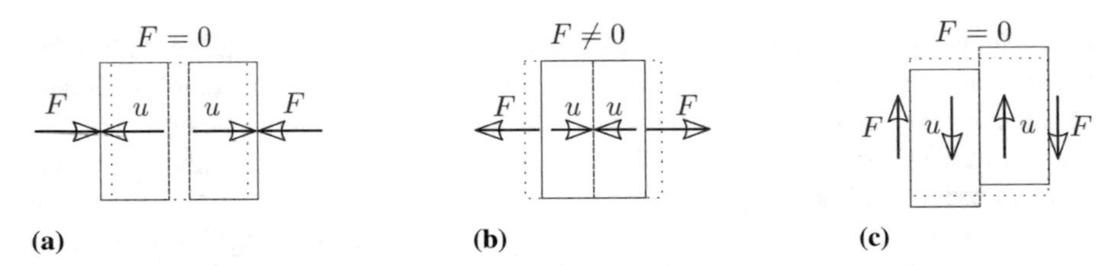

Fig. 10.15 Basic characteristics of an ideal plane and frictionless crack: (a) crack opening without resistance, (b) transmission of contact forces and (c) sliding of the crack faces against each other without resistance

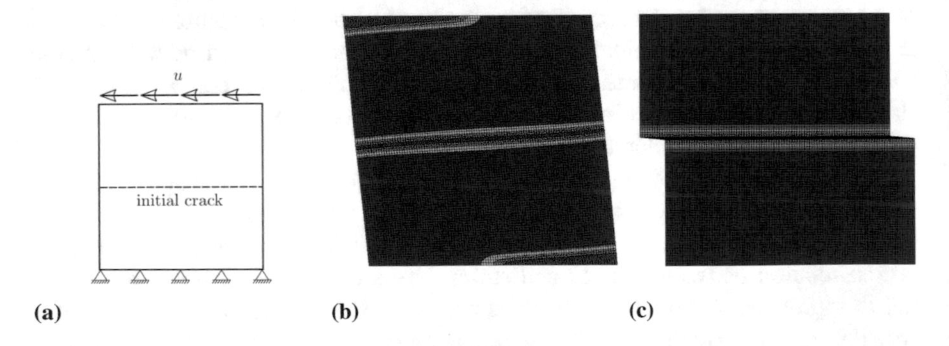

Fig. 10.16 Benchmark simulation for the sliding of crack faces against each other without resistance: (a) setup and boundary conditions, (b) spectral decomposition and (c) directional decomposition

$$\underbrace{\begin{bmatrix} -0.3 & 0.0 & 0.0 \\ 0.0 & 0.0 & 0.0 \\ 0.0 & 0.0 & 0.0 \end{bmatrix}}_{\varepsilon} = \underbrace{\begin{bmatrix} -0.1 & 0.0 & 0.0 \\ 0.0 & -0.1 & 0.0 \\ 0.0 & 0.0 & -0.1 \end{bmatrix}}_{\varepsilon_{\mathrm{V}}} + \underbrace{\begin{bmatrix} -0.2 & 0.0 & 0.0 \\ 0.0 & 0.1 & 0.0 \\ 0.0 & 0.0 & 0.1 \end{bmatrix}}_{\varepsilon_{\mathrm{D}}}.$$

Given a crack orientation such, that the uniaxial compressive strain is perpendicular on the crack surface, the total stress should be undegraded, because it is actually transmitted via contact in reality. In contrast, a significant amount is considered to belong to the deviatoric strain component and, thus, is degraded in the numerical model. This problem is amplified at the edge of structures, see Fig. 10.17b. The degradation of the deviatoric component results in a reduction of the stiffness against the lateral expansion. Without the stabilizing effect of neighboring nodes, the lateral deformation of the edge nodes of the discretization is not bounded any more and leads to instability and non-convergent results for the numerical simulation. In contrast, the directional split enables the simulation of a realistic post-fracture behavior in this case, too.

10.5 Conclusions

A novel framework is presented to model the complex structural behavior of SHCC. SHCC is a composite material that consists of a fine grained cement matrix reinforced by high performance polymer fibers. Tensile loading on the structural level results in three stages of characteristic behavior. The initial elastic response is followed by a relatively long stage of micro crack evolution, where each micro crack is bridged by the polymer fibers and the structural integrity and load bearing capacity is preserved. Finally, structural failure is a result of pull out or rupture of fibers in a critical micro

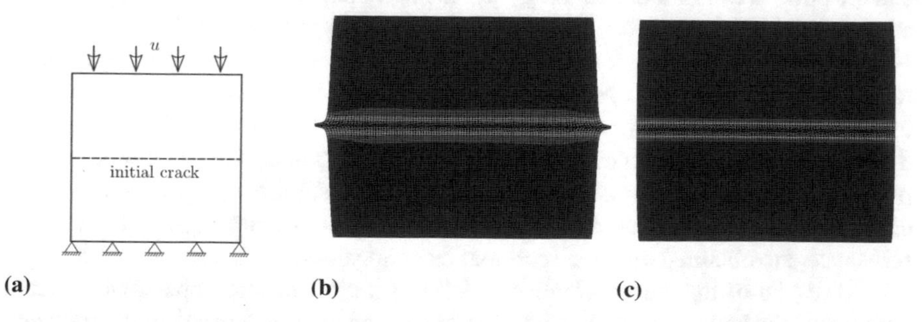

Fig. 10.17 Benchmark simulation for the transmission of contact forces perpendicular to the crack faces: (a) setup and boundary conditions, (b) volumetric-deviatoric decomposition and (c) directional decomposition

crack. The framework proposed here is to represent this behavior by an elasto-plastic material model combined with damage or fracture to approximate pull out or rupture, respectively.

The theory of the gradient enhanced microplane model for damage and plasticity is recapitulated. A Drucker-Prager yield function with caps for tension and compression is specified and evaluated with the undamaged stresses at the level of the microplanes. The weighted average of the local and the non-local equivalent strain governs the evolution of damage.

The simulations with the microplane approach show a good agreement to the experimental observations. The uniaxial compressive strength is calibrated in order to fit the transition from linear elasticity to the evolution of multiple micro cracks, where the structural response is approximated by isotropic hardening. Based on the calibration of the threshold for tensile damage, the softening observed after the ultimate strain is represented by damage evolution. A functional correlation in terms of a DIF is proposed between the strain rate and the calibrated model parameters based on the quasi-static simulations.

The phase-field model for ductile fracture with von Mises yield function and volumetric-deviatoric split of the strain energy density is applied. The transition from brittle to ductile fracture is obtained by modification of the degradation function with respect to the equivalent plastic strain. Furthermore, the directional decomposition of the stresses for a realistic approximation of the post-fracture behavior is presented.

In general, the global characteristic of the structural response are obtained by the phase-field model. The initial yield strength is calibrated to fit the transition from linear elasticity to evolution of multiple micro cracks. However, the linear hardening modulus obtained with the model is not suitable to represent the experimental results properly. Instead, non-linear hardening should be used. The final structural failure is modeled by the evolution of a phase-field crack at the ultimate strain and exhibits the experimentally observed abrupt drop of the reaction force. Here, two model parameters of the phase-field model, i.e. the critical equivalent plastic strain and the fracture toughness, have concurrent impact on the evolution of the phase-field crack. Two sets of calibration are performed, where either of these two parameters is fixed and the other is calibrated. In analogy to the microplane results, three DIF functions are proposed to relate the calibrated model parameters with the strain rate.

An additional benchmark simulation on a structure with an existing crack is used to demonstrate the capability of the directional decomposition of the stresses in terms of a realistic approximation of a post-fracture behavior. In contrast to the volumetric-deviatoric and the spectral decomposition, the three characteristic features of an ideal plane crack without friction, i.e. transmission of contact force perpendicular to the crack face and the separation and sliding of the crack faces against each other without resistance, are obtained by the directional decomposition

The results of the study establish a basis to formulate rate dependent material parameters for both models. A validation of such an approach requires the transient simulation of the experiment with different loading rates. Additional experimental evidence to understand the transition between the two ways of structural failure, i.e. pull out or rupture of fibers, can be incorporated into a combination of the microplane model and the phase-field approach in order to provide an all-embracing modeling approach for SHCC. To this end, the phase-field model needs to be enhanced by non-linear hardening and the directional split in addition.

Acknowledgements The authors would like to acknowledge the financial support of "German Research Foundation" under Grant KA 1163/19 and of ANSYS Inc., Canonsburg, USA and as well the technical support of the center for information services and high performance computing of TU Dresden for providing access to the Bull HPC-Cluster.

References

Ambati M, Gerasimov T, De Lorenzis L (2015) Phase-field modeling of ductile fracture. Computational Mechanics 55(5):1017–1040, DOI 10.1007/s00466-015-1151-4

Amor H, Marigo JJ, Maurini C (2009) Regularized formulation of the variational brittle fracture with unilateral contact: Numerical experiments. Journal of the Mechanics and Physics of Solids 57(8):1209 – 1229, DOI 10.1016/j.jmps.2009.04.011

Curosu I (2017) Influence of fiber type and matrix composition on the tensile behavior of strain-hardening cement-based composites (shcc) under impact loading. Phd thesis, TU Dresden, Dresden

Hofacker M, Welschinger F, Miehe C (2009) A variational–based formulation of regularized brittle fracture. PAMM 9(1):207–208, DOI 10.1002/pamm.200910078

Loew PJ, Peters B, Beex LAA (2019) Rate-dependent phase-field damage modeling of rubber and its experimental parameter identification. Journal of the Mechanics and Physics of Solids 127:266 – 294, DOI 10.1016/j.jmps.2019.03.022

Steinke C, Kaliske M (2018) A phase-field crack model based on directional stress decomposition. Computational Mechanics 63(5):1019–1046, DOI 10.1007/s00466-018-1635-0

Steinke C, Özenç K, Chinaryan G, Kaliske M (2016) A comparative study of the r-adaptive material force approach and the phase-field method in dynamic fracture. International Journal of Fracture 201(1):97–118, DOI 10.1007/s10704-016-0125-7

Zreid I, Kaliske M (2018) A gradient enhanced plasticity–damage microplane model for concrete. Computational Mechanics 62(5):1239–1257, DOI 10.1007/s00466-018-1561-1

Chapter 11
Effective Properties of Composite Material Based on Total Strain Energy Equivalence

Anna Wiśniewska, Szymon Hernik, and Halina Egner

Abstract In the present work the mechanical equivalence hypothesis, classically used in continuum damage mechanics problems, was applied to estimate the elasto-plastic properties of isotropic composite materials. The equivalence of total internal energy was postulated between a real, heterogeneous composite material, and a fictitious, quasi-homogeneous configuration. The properties of a composite material were expressed as analytical functions of an inclusion volume fraction and properties of constituent materials. The results were compared with the results of several other methods of effective elastic properties estimation. In the inelastic range of the material response the proposed approach was examined by means of parametric studies to show its ability to reflect different experimentally observed features of real composite materials.

Key words: Effective properties · Composite material · Constitutive modeling

11.1 Introduction

A composite material is often characterized by a multiphase microstructure where each phase exhibits different mechanical properties. In the present work we consider a material that consists of two isotropic phases. A new method of estimating the effective properties of such composite in the macro-scale is derived, based on a mechanical equivalence hypothesis. The concept of mechanical equivalence between the real and fictitious material configurations, classically used in continuum damage

Anna Wiśniewska · Szymon Hernik · Halina Egner

Institute of Machine Design, Faculty of Mechanical Engineering, Cracow University of Technology, Al. Jana Pawła II 37, 31-864 Kraków, Poland
e-mail: anna.wisniewska1@pk.edu.pl, hernik@mech.pk.edu.pl, halina.egner@pk.edu.pl

H. Altenbach et al. (eds.), *Plasticity, Damage and Fracture in Advanced Materials*, Advanced Structured Materials 121,
https://doi.org/10.1007/978-3-030-34851-9_11

mechanics problems (Murakami, 2012) was recently extended to a general multi-dissipative material modelling by Egner and Ryś (2017); Ryś and Egner (2019). One of the advantages of such approach is the possibility to describe not only the elastic behavior, but also the general irreversible behavior in a consistent manner. The results presented here concern the case of a particle-reinforced composite in a disordered way (isotropic symmetry). However, the method is general and can be applied also to more complex cases.

Continuum mechanics approach applied in the present paper provides the constitutive modeling in the framework of thermodynamics of irreversible processes with internal state variables. This approach is based on a concept of the effective quasi-homogeneous continuum (see Fig. 11.1). The material heterogeneity (on the micro and mesoscale) is smeared out over the representative volume element (RVE) of the heterogeneous material. The true state of a material within RVE, is mapped to a material point of the effective quasi-homogeneous continuum. The true distribution of micro-structures within the RVE, is reflected by the change of the macroscopic constitutive properties. The microstructural rearrangements are defined by the set of internal state variables Λ_k of the scalar, vectorial, or tensorial nature (Chaboche, 1997; Egner, 2012; Ganczarski et al, 2010; Murakami and Ohno, 1981; Skrzypek and Kuna-Ciskał, 2003). In the presented approach the constitutive tensors for the composite material are defined by the use of even-rank effect tensors that map

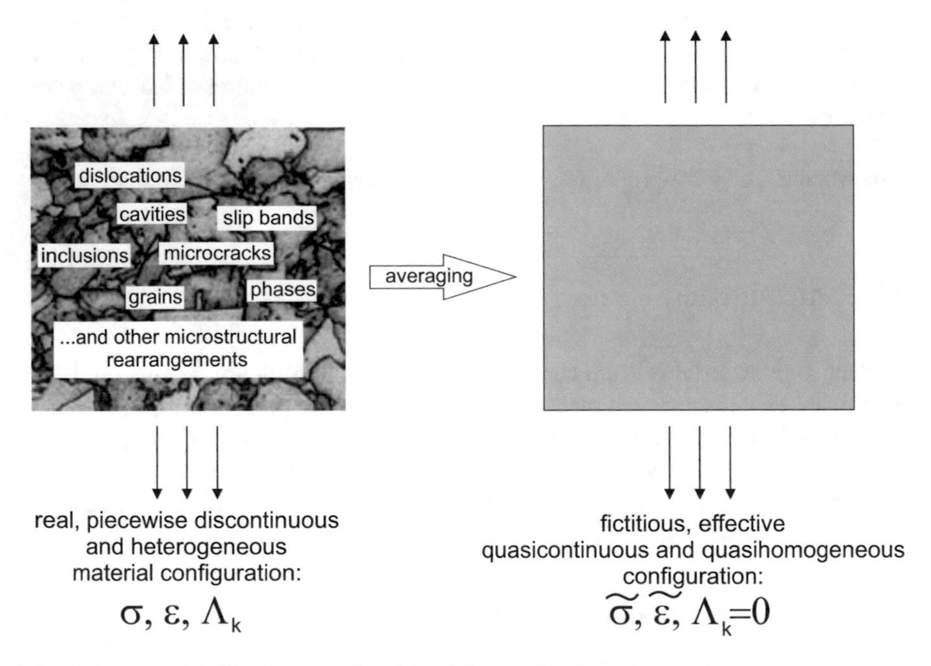

Fig. 11.1 Physical (multi-phase), and equivalent (mono-phase) continuum; the total strain energy equivalence principle is used in order to smear out the true inclusion, disclocations, etc. distribution over the RVE to yield the effective constitutive moduli for a heterogeneous material

thermodynamic forces from real multi-component (heterogeneous) to the fictitious (pseudo-homogeneous) configuration.

11.2 Basic Relations

11.2.1 State Variables

In the approach based on thermodynamics of irreversible processes with internal state variables the state of a material is entirely determined by certain values of some independent variables, called variables of state. For elastic-plastic material exhibiting mixed hardening, in the isothermal conditions, the complete set of state variables $\{V_{st}\}$ reflecting the current state of the thermodynamic system consists of observable variables: elastic (or total) strain tensor ε_{ij}^{e} and hardening state variables:

$$\{V_{st}\} = \{\varepsilon_{ij}^{e}, r, \alpha_{ij}\} \tag{11.1}$$

To describe the influence of reinforcement on the composite properties the scalar parameter

$$\xi = \frac{\mathrm{d}V^{\mathrm{inc}}}{\mathrm{d}V^{\mathrm{RVE}}} \tag{11.2}$$

will be here used, which denotes the volume fraction of the reinforcing inclusions in the RVE. However, this parameter is not a state variable that undergoes evolution.

11.2.2 Effective State Variables

The definition of effective state variables is related to the hypothesis of the mechanical equivalence between a real material configuration and the fictitious one. Various equivalence hypotheses have been formulated so far, for example strain equivalence, stress equivalence, strain energy or complementary energy equivalence hypotheses, or finally, total energy equivalence hypothesis (Chow and Lu, 1992; Saanouni et al, 1994). According to these hypotheses, the effective state variables are defined in such a way that respectively strains, stresses, strain energy or complementary strain energy, or total energy for both real and fictitious materials are the same.

The total energy equivalence hypothesis (H1) applied in the present research we formulate in the following way (Wiśniewska et al, 2019):

H1: *At any time (t), to an RVE in its real (deformed, multiphase, etc.) configuration, described by the set of state variable pairs, we associate an unchanged monophase equivalent fictive configuration, the state of which is described by the effective state variables – in such a manner that the total internal energy defined over the two (real and fictive) configurations is the same.*

Let us consider a real, heterogeneous configuration (R) of a composite material. We introduce a fictitious quasi-homogeneous configuration (F), characterized by the couples of effective state variables and effective thermodynamic forces defined on the basis of the total energy equivalence hypothesis (H1). Mapping from configuration (R) to (F) may be performed with the use of a new inclusion-effect tensor N_{ijkl}, defined in analogy to the classical damage effect tensor (Egner and Ryś, 2017; Ryś and Egner, 2019).

The general expression for the equivalence of total internal energy between subsequent configurations may be written in the following way:

$$U_t\left(\varepsilon_{ij}^e, \alpha_{ij}, r; \xi\right) = \tilde{U}_t\left(\tilde{\varepsilon}_{ij}^e, \tilde{\alpha}_{ij}, \tilde{r}; 0\right) \tag{11.3}$$

Hypothesis (H1) expressed by Eq. (11.3) will be further expressed in a stronger form, that all the energy components: the reversible elastic energy U_E, kinematic hardening energy U_{kin} and isotropic hardening energy U_{iso}, are equivalent:

$$U_E = \frac{1}{2}\sigma_{ij}\varepsilon_{ij}^e = \frac{1}{2}\tilde{\sigma}_{ij}\tilde{\varepsilon}_{ij}^e \tag{11.4}$$

$$U_{kin} = \frac{1}{2}X_{ij}\alpha_{ij} = \frac{1}{2}\tilde{X}_{ij}\tilde{\alpha}_{ij} \tag{11.5}$$

$$U_{iso} = \frac{1}{2}Rr = \frac{1}{2}\tilde{R}\tilde{r} \tag{11.6}$$

In the above equations σ_{ij} denotes the stress tensor, while X_{ij} and R stand for the back stress and drag stress thermodynamic forces, conjugated to plastic hardening variables:

$$X_{ij} = C(\xi)\alpha_{ij} \tag{11.7}$$

$$R = Q(\xi)r \tag{11.8}$$

A general solution that satisfies equation (11.4) may take the following form:

$$\tilde{\sigma}_{ij} = \left[N_{ijkl}^e(\xi)\right]^{-1} \sigma_{kl} \tag{11.9}$$

$$\tilde{\varepsilon}_{ij}^e = \left[N_{ijkl}^e(\xi)\right]^{T} \varepsilon_{kl}^e \tag{11.10}$$

In the above equations $N_{ijkl}^e(\xi)$ is a symmetric fourth-order operator function of variable ξ. This operator should exhibit certain characteristics. Namely, tensor $N_{ijkl}^e(\xi)$ should:

- be positive definite, symmetric and monotonic function of variable ξ;
- be reduced to the fourth-rank unit tensor in the absence of inclusions, $\xi = 0$;
- transform the properties of the matrix material into the properties of the inclusion material when the variable ξ reaches unity.

For isotropic composite we assume that the inclusion effect tensor has the isotropic two-parameter form (Wiśniewska et al, 2019):

$$N^{e}_{ijkl}(\xi) = f_1(\xi)\delta_{ij}\delta_{kl} + f_2(\xi)\left(\delta_{ik}\delta_{jl} + \delta_{il}\delta_{jk}\right) \tag{11.11}$$

where $f_1(\xi)$ and $f_2(\xi)$ are scalar valued functions of the volume fraction of inclusions.

In analogy to (11.9)–(11.10) a general solution that satisfies equation (11.5) may take the following form:

$$\tilde{X}_{ij} = \left[N^{p}_{ijpq}(\xi)\right]^{-1} X_{pq}, \qquad \tilde{\alpha}_{ij} = \left[N^{p}_{ijpq}(\xi)\right]^{T} \alpha_{pq} \tag{11.12}$$

"Kinematic" operator $N^{p}_{ijpq}(\xi)$ should exhibit the same features as "elastic" operator (see above), but for simplicity in the present considerations we define it in the simplest form of a unimodular fourth-rank tensor:

$$N^{p}_{ijpq} = h^{kin}(\xi)\mathbf{I}_{ijpq} \tag{11.13}$$

To satisfy equation (11.6) we adopt here the following relations:

$$\tilde{R} = \frac{R}{h^{iso}(\xi)}, \qquad \tilde{r} = h^{iso}(\xi)r \tag{11.14}$$

In Eqs (11.13) and (11.14) $h^{iso}(\xi)$ and $h^{kin}(\xi)$ denote scalar-valued function of parameter ξ.

11.3 Effective Composite Properties

11.3.1 Elastic Range

By substituting Eqs. (11.9)-(11.10) into Eq. (11.4), and applying Hooke's law

$$\sigma_{ij} = E_{ijkl}\varepsilon^{e}_{kl}$$

one can express the inclusion affected elasticity tensor of a real RVE material, $E_{ijkl}(\xi)$, in terms of the corresponding elasticity tensor of a matrix material, E^{0}_{ijkl}, by the following relation:

$$E_{ijkl}(\xi) = N^{e}_{ijpq}(\xi)E^{0}_{pqrs}N^{e}_{rskl}(\xi) \tag{11.15}$$

Functions $f_1(\xi)$ and $f_2(\xi)$ appearing in Eq. (11.11) may be determined by applying boundary conditions related to the inclusion content. For $\xi = 0$ (no inclusions) the elastic properties (11.15) remain unaffected by parameter ξ. On the other hand, when $\xi = 1$, the elastic properties become equal to the properties of the inclusion material. The second boundary condition (100% of inclusions) is rather theoretical, while in technical applications the inclusion content is much lower. With the use of Eq. (11.11), applying proper boundary conditions for $\xi = 0$ (no inclusions) and $\xi = 1$ (100% of inclusions) relations (11.11) and (11.15) may be rewritten in the following form (Wiśniewska et al, 2019):

$$
\begin{aligned}
N^{e}_{ijkl}(\xi) = &-\frac{1}{3}\left(\sqrt{\frac{\mu^1}{\mu^0}} - \sqrt{\frac{K^1}{K^0}}\right)\xi\delta_{ij}\delta_{kl} \\
&+ \left[\frac{1}{2}\left(\sqrt{\frac{\mu^1}{\mu^0}} - 1\right)\xi + \frac{1}{2}\right]\left(\delta_{ik}\delta_{jl} + \delta_{il}\delta_{jk}\right)
\end{aligned} \tag{11.16}
$$

$$
E^{e}_{ijkl}(\xi) = \lambda(\xi)\delta_{ij}\delta_{kl} + \mu(\xi)\left(\delta_{ik}\delta_{jl} + \delta_{il}\delta_{jk}\right) \tag{11.17}
$$

where $\lambda(\xi)$ and $\mu(\xi)$ are the effective Lamé material characteristics, dependent on the volume fraction of inclusions:

$$
\begin{aligned}
\lambda(\xi) = \lambda^0 &\left\{2\left(\sqrt{\frac{\mu^1}{\mu^0}} - \sqrt{\frac{K^1}{K^0}}\right)\left[\left(1 - \frac{1}{2}\left(\sqrt{\frac{\mu^1}{\mu^0}} + \sqrt{\frac{K^1}{K^0}}\right)\right)\xi^2 - \xi\right]\right. \\
&+ \left.\left[\left(\sqrt{\frac{\mu^1}{\mu^0}} - 1\right)\xi + 1\right]^2\right\} \\
&+ \frac{4}{3}\mu^0\left(\sqrt{\frac{\mu^1}{\mu^0}} - \sqrt{\frac{K^1}{K^0}}\right)\left[\left(1 - \frac{1}{2}\left(\sqrt{\frac{\mu^1}{\mu^0}} + \sqrt{\frac{K^1}{K^0}}\right)\right)\xi^2 - \xi\right]
\end{aligned} \tag{11.18}
$$

$$
\mu(\xi) = \mu^0\left[\left(\sqrt{\frac{\mu^1}{\mu^0}} - 1\right)\xi + 1\right]^2 \tag{11.19}
$$

In the above relations (μ^1, λ^1) and (μ^0, λ^0) are Lamé parameters, while

$$
K^1 = \lambda^1 + \frac{2}{3}\mu^1
$$

and

$$
K^0 = \lambda^0 + \frac{2}{3}\mu^0
$$

are the Helmholtz coefficients, of inclusion and matrix materials, respectively.

According to Eqs. (11.9) and (11.16) the stress tensor in the fictitious equivalent configuration takes the form:

$$
\tilde{\sigma}_{ij} = \frac{1}{2f_2}\sigma_{ij} - \frac{f_1}{2f_2\left(3f_1 + 2f_2\right)}\sigma_{kk}\delta_{ij} \tag{11.20}
$$

where

$$
f_1(\xi) = -\frac{1}{3}\left(\sqrt{\frac{\mu^1}{\mu^0}} - \sqrt{\frac{K^1}{K^0}}\right)\xi, \qquad f_2(\xi) = \frac{1}{2}\left(\sqrt{\frac{\mu^1}{\mu^0}} - 1\right)\xi + \frac{1}{2} \tag{11.21}
$$

11.3.2 Plastic Range

The yield stress $\sigma_y(\xi)$ has to fulfill boundary conditions in two characteristic points, namely $\xi = 0$ (matrix material, characterized by yield stress σ_y^0), and $\xi = 1$ (inclusion material with yield stress σ_y^1):

$$\sigma_y(\xi = 0) = \sigma_y^0, \qquad \sigma_y(\xi = 1) = \sigma_y^1 \tag{11.22}$$

The simplest linear approximation between the above points gives:

$$\sigma_y(\xi) = \left(\sigma_y^1 - \sigma_y^0\right)\xi + \sigma_y^0 \tag{11.23}$$

According to hypothesis (H1) and taking into account Eqs. (11.5)-(11.6) together with the assumptions (11.12) and (11.13)-(11.14), the forces conjugated to hardening state variables have the form:

$$X_{ij} = \left[h^{kin}(\xi)\right]^2 C^0 \alpha_{ij} \tag{11.24}$$

$$R = \left[h^{iso}(\xi)\right]^2 Q^0 r \tag{11.25}$$

Comparison of Eq. (11.24) with (11.7), and (11.25) with (11.8) results in:

$$C(\xi) = \left[h^{kin}(\xi)\right]^2 C^0 \tag{11.26}$$

$$Q(\xi) = \left[h^{iso}(\xi)\right]^2 Q^0 \tag{11.27}$$

The effective isotropic composite properties in the whole range of elastic-plastic behavior may therefore be expressed with the use of constituent properties as functions of the inclusion volume fraction ξ.

11.4 Validation

11.4.1 Elastic Properties

The homogenization procedure was verified on the basis of experimental data and compared with other estimation methods. The analysis has been carried out for several composites. One of them was the glass short-fiber reinforced polyamide. The material data of polyamide and glass short-fiber are presented in Table 11.1.

Figure 11.2 shows estimations of Young's modulus as a function of the particle volume fraction according to three independent methods. One of them (E_{eff}) is the energy equivalence based method proposed in the presented paper, while the other two are Voigt (E^V) Voigt (1889) and Reuss (E^R) Reuss (1929) rules. Proposed

Table 11.1 Material properties of matrix polyamide and glass short-fiber

Material	Young modulus E [GPa]	Poisson ratio ν [-]	Lamé parameter λ [GPa]	Lamé parameter μ [GPa]
PA	3.2*	0.39**	11.74	1.15
glass short-fiber	72.0**	0.21**	102.89	29.75

* – own data
** – adopted from database `http://matweb.com`

method is reflected by the curve placed between the Voigt and Reuss lower and upper bounds.

Similar comparisons of experimental data with theoretical estimations are presented in Fig. 11.3, in which the distribution of Young modulus for hydroxyapatite reinforced PE is shown. The material data of PE and HAp are presented in Table 11.2.

Presented model (EE) was also compared with other micromechanical and numerical approaches: the self-consistent method (SC), the Hashin-Shtrikman upper

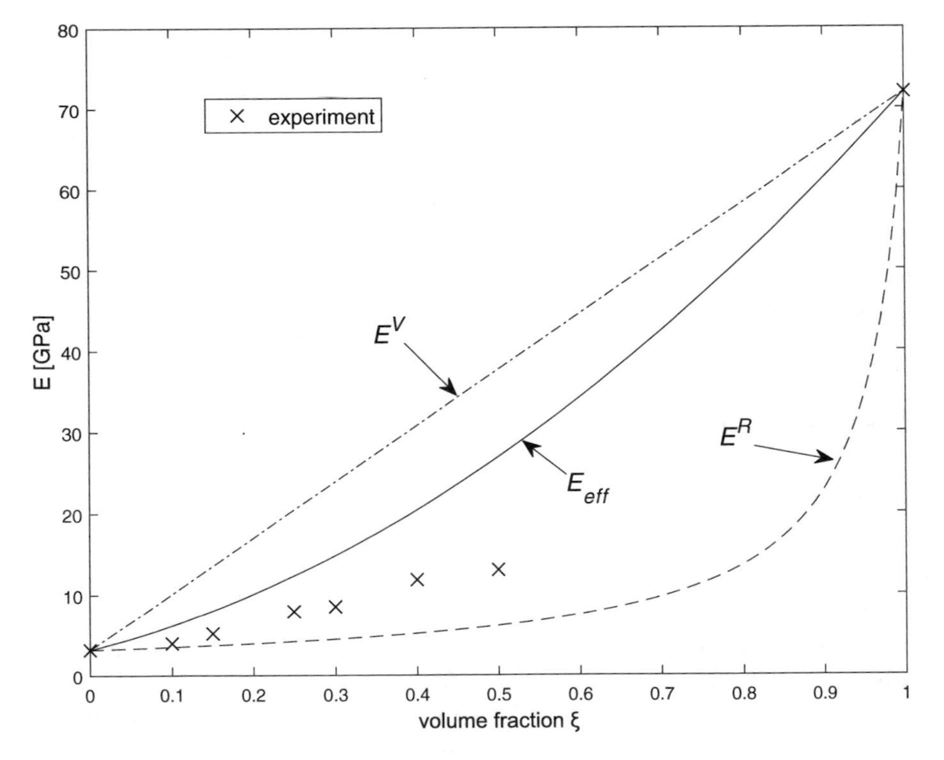

Fig. 11.2 Young's modulus distribution obtained from different estimation methods: E_{eff} (method proposed in the presented research), E^V (the Voigt estimation), E^R (the Reuss estimation) for glass short-fiber reinforced polyamide. Data adapted from Koszkul (2001)

Table 11.2 Material properties of matrix PE and inclusions HAp

Material	Young modulus E [GPa]	Poisson ratio ν [-]	Lamé parameter λ [GPa]	Lamé parameter μ [GPa]
PE	1.30*	0.40	1.86	0.46
HAp	13.50**	0.14**	2.72	7.00

* – adopted from Bonfield (1988)
** – adopted from Charriére et al (2001)

(SH) and lower (HS) bounds, the generalized self-consistent method (GSC), the composite sphere approach (CS), the composite sphere upper bound obtained under displacement boundary conditions (CSu), and the finite element estimates (FE-S). Analysis was made on the basis of results presented in Kursa et al (2018). The comparison was performed for the composite material AA6061-SiC (see Fig. 11.4). The material data for the phases (matrix and inclusions) are given in Table 11.3. The

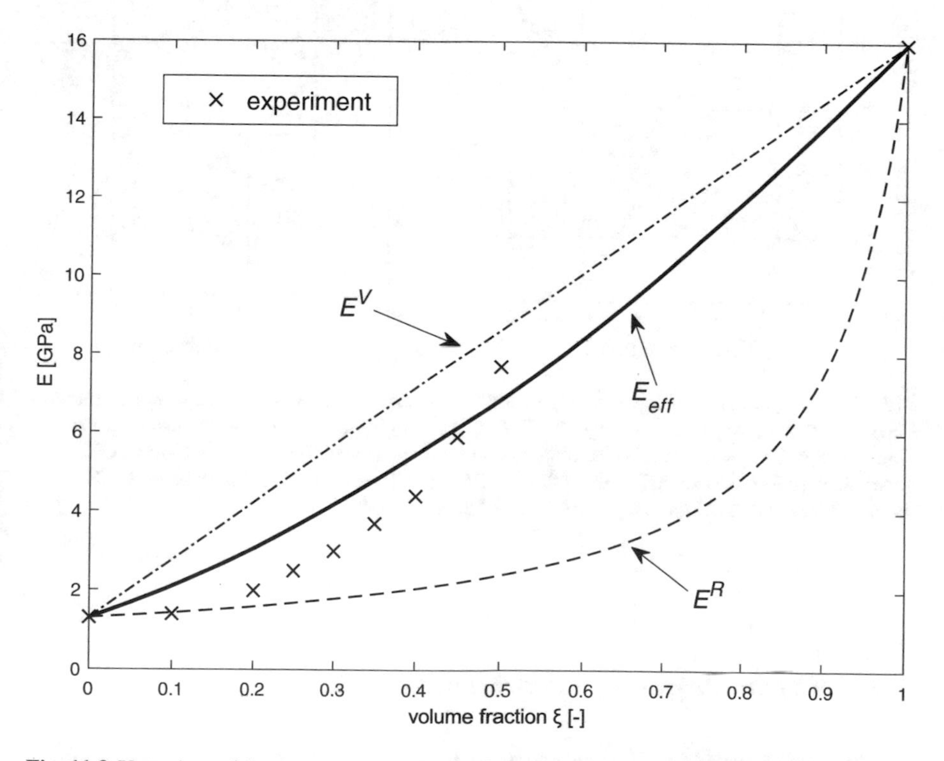

Fig. 11.3 Young's modulus versus volume fraction of reinforcement obtained from different estimation methods: E_{eff} (method proposed in the presented research), E^V (the Voigt estimation), E^R (the Reuss estimation) for hydroxyapatite reinforced PE. Data adapted from Charriére et al (2001) and Bonfield (1988)

results obtained by the method proposed in the present paper are close to the upper Hashin-Shtrikman bound.

Table 11.3 Material properties of matrix AA6061 and inclusions SiC

Material	Young modulus E [GPa]	Poisson ratio ν [-]	Lamé parameter λ [GPa]	Lamé parameter μ [GPa]
AA6061	70.0*	0.33	51.08	26.31
SiC	450.0**	0.17**	99.07	192.31

* – adopted from ASM (1990)
** – adopted from `http://www.intlceramics.com` (Last accessed: 24th February 2016)

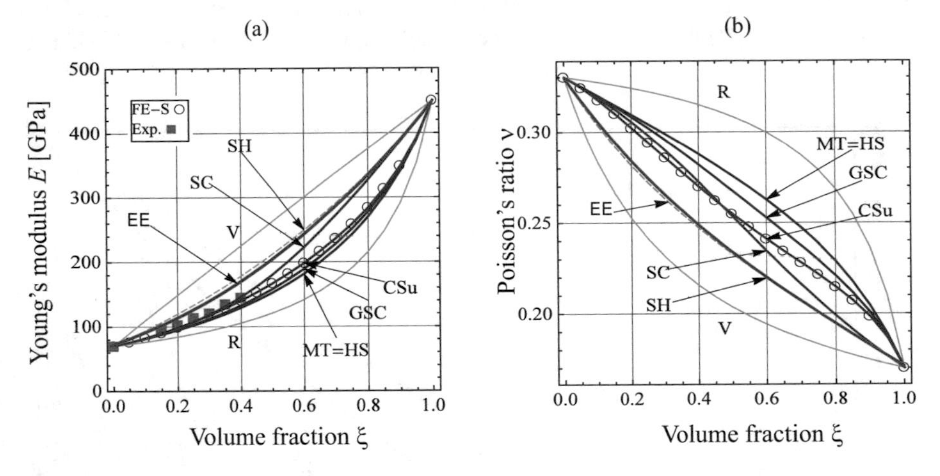

Fig. 11.4 Comparison of effective Young's modulus obtained with the use of the proposed method (EE) and other approaches: V – Voigt, R – Reuss, SC – self-consistent, CS – composite sphere, CSu – composite sphere upper bound obtained under displacement boundary conditions, GSC – generalized self-consistent, MT – Mori-Tanaka, SH and HS – Hashin-Shtrikman bounds, EE – method developed in the present paper (after Kursa et al, 2018)

11.4.2 Plastic Properties - Parametric Studies

To illustrate the results of the proposed approach for plastic properties, the following hardening functions were proposed:

$$h^{kin}(\xi) = \sqrt{h_X\xi^2 + \left(\frac{C^1}{C^0} - h_{X-1}\right)\xi + 1} \tag{11.28}$$

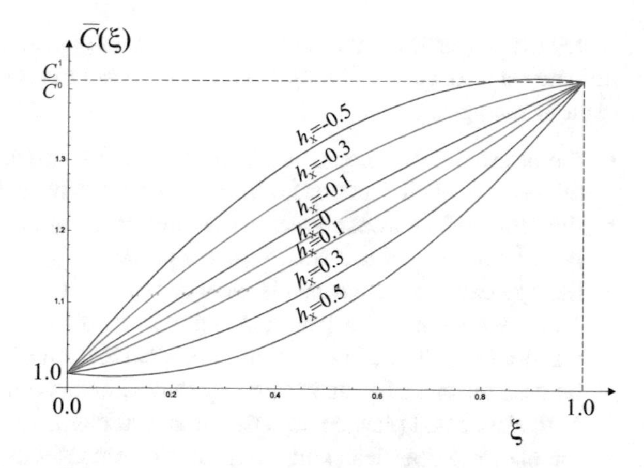

Fig. 11.5 Influence of h_X parameter on plastic modulus $C(\xi)$.

and

$$h^{iso}(\xi) = \sqrt{h_R\xi^2 + \left(\frac{Q^1}{Q^0} - h_{R-1}\right)\xi + 1} \tag{11.29}$$

Such functions fulfill the boundary conditions for the pure matrix material ($\xi = 0$):

$$C(\xi = 0) = C^0, \qquad Q(\xi = 0) = Q^0 \tag{11.30}$$

And for the pure inclusion material, $\xi = 1$:

$$C(\xi = 1) = C^1, \qquad Q(\xi = 1) = Q^1 \tag{11.31}$$

The admissible values of material parameters h_x and h_R should ensure that hardening functions $h^{kin}(\xi)$ and $h^{iso}(\xi)$ are real valued and monotonic increasing. For this reason parameters h_x and h_R are placed between their lower and upper bounds:

$$|h_X| \leq \frac{C^1}{C^0}, \qquad |h_R| \leq \frac{Q^1}{Q^0} \tag{11.32}$$

Hardening functions (11.28) and (11.29) allow to reflect the influence of the reinforcement on the plastic characteristics of the composite material for a wide range of behaviors. The results for hardening modulus $C(\xi)$ are presented in Fig. 11.5.

11.5 Conclusions

In the present paper the approach based on the total energy equivalence hypothesis, originally developed for damaged materials, was extended to composite materials modeling. The elastic-plastic isotropic composite material was considered, and its properties were estimated as analytical functions of the inclusion volume fraction, in which the properties of the constituent materials appear as parameters. The proposed here new approach to estimating material characteristics of composite materials has clear advantages:

- the properties of composite materials at a macroscale can unequivocally be obtained as explicit functions of reinforcement volume fraction,
- the proposed approach assures the correct Voigt and Reuss boundary values of both Young's modulus and Poisson's ratio,
- the approach has a reasonable, energy-based physical interpretation,
- it can be used both for isotropic and anisotropic composite materials (however, a tensorial measure of the amount and distribution of reinforcements should then be used instead of a scalar quantity that accounts only for the volume fraction),
- both elastic and plastic composite material characteristics can be obtained,
- the elastic properties exhibit a good agreement with the available experimental data and with the results of other homogenization techniques.

On the other hand, there are some aspect that should be solved and developed:

- The mathematical form of the elastic influence effect tensor (11.16) allows to model the effect of inclusions on both elastic characteristics, Young's modulus $E(\xi)$ and Poisson's ratio $\nu(\xi)$. However, at the same time such form of tensor $N^e_{ijkl}(\xi)$ leads to the effective stress definition (11.20), which results in a tri-axial effective stress state in a uniaxial tension/compression. This can be avoided by simplifying the definition of $N^e_{ijkl}(\xi)$, but then only Young's modulus will be influenced by the volume fraction ξ.
- In the present version the model does not regard for the influence of the particle size.
- It is assumed here that the particle-reinforced composites in a disordered manner show isotropic symmetry after homogenization (at the level of RVE). However, it may happen that the averaged material moduli at the macro-scale (composite level) exhibit other than isotropic symmetry properties. In such case the presented approach should be extended by the use of proper tensorial measures of the reinforcement influence.

Acknowledgements The Grant UMO-2018/31/N/ST8/01052 from the National Science Center NCN, Poland is gratefully acknowledged.

References

ASM (1990) ASM Handbook, vol 2 Properties and Selection: Nonferrous Alloys and Special-purpose Materials. ASM International Handbook Committee

Bonfield W (1988) Hydroxyapatite-reinforced polyethylene as an analogous material for bone replacement. Annals of the New York Academy of Sciences 523:173–177

Chaboche J (1997) Thermodynamic formulation of constitutive equations and application to the viscoplasticity and viscoelasticity of metals and polymers. International Journal of Solids and Structures 34(18):2239–2254

Charriére E, Terrazzoni S, Pittet C, Mordasini P, Dutoit M, Lemaître J, Zysset P (2001) Mechanical characterization of brushite and hydroxyapatite cements. Biomaterials 22:2937–2945

Chow C, Lu T (1992) An analytical and experimental study of mixed-mode ductile fracture under nonproportional loading. International Journal of Damage Mechanics 1:191–236

Egner H (2012) On the full coupling between thermo-plasticity and thermo-damage in thermodynamic modeling of dissipative materials. International Journal of Solids and Structures 49(2):279–288

Egner H, Ryś M (2017) Total energy equivalence in constitutive modelling of multidissipative materials. International Journal of Damage Mechanics 3:417–446

Ganczarski AW, Egner H, Muc A, Skrzypek JJ (2010) Constitutive models for analysis and design of multifunctional technological materials. In: Rustichelli F, Skrzypek J (eds) Innovative Technological Materials: Structural Properties by Neutron Scattering, Synchrotron Radiation and Modeling, Springer, pp 179–223

Koszkul J (2001) Kompozyty poliamidu 6 z włóknem szklanym. Composites 1(2):159–162

Kursa M, Kowalczyk-Gajewska K, Lewandowski MJ, Petryk H (2018) Elastic-plastic properties of metal matrix composites: validation of meanfeld approaches. European Journal of Mechanics - A/Solids 68:53–66

Murakami S (2012) Continuum damage mechanics: A continuum mechanics approach to the analysis of damage and fracture. Springer-Verlag

Murakami S, Ohno N (1981) A continuum theory of creep and creep damage. In: Ponter A, Hayhurst D (eds) Creep in Structures, Springer, Berlin, 3rd IUTAM Symposium on Creep in Structures, pp 422–444

Reuss A (1929) Berechnung der Fließgrenze von Mischkristallen auf Grund der Plastizitätsbedingung für Einkristalle. Zeitschrift für angewandte Mathematik und Mechanik 9(1):49–58

Ryś M, Egner H (2019) Energy equivalence based constitutive model of austenitic stainless steel at cryogenic temperatures. International Journal of Solids and Structures 164:52–65

Saanouni K, Forster C, Ben Hatira F (1994) On the inelastic flow with damage. International Journal of Damage Mechanics 3:140–169

Skrzypek JJ, Kuna-Ciskał H (2003) Anisotropic elastic-brittle-damage and fracture models based on irreversible thermodynamics. In: Skrzypek JJ, Ganczarski AW (eds) Anisotropic Behaviour of Damaged Materials, Springer, Berlin, Heidelberg, Lecture Notes in Applied and Computational Mechanics, vol 9, pp 143–184

Voigt W (1889) Über die Beziehungen zwischen den beiden Elastizitätskonstanten isotroper Körper. Annalen der Physik 274(12):573–587

Wiśniewska A, Hernik S, Liber-Kneć A, Egner H (2019) Effective properties of composite material based on total strain energy equivalence. Composites Part B Engineering 166:213–220

Printed in the United States
By Bookmasters